MAPPING THE WORLD

MAPS AND THEIR HISTORY

MAPPING THE WORLD

MAPS AND THEIR HISTORY

Nathaniel Harris

Brown
PARTWORKS

© 2002 Brown Partworks Limited
This edition published in the UK by Bookmart Limited

ISBN 1 84044 087 2

Produced by Brown Partworks Limited
8 Chapel Place, Rivington Street, London EC2A 3DQ
www.brownpartworks.co.uk

Managing Editor: Tim Cooke
Editor: Dawn Titmus
Designer: Ian Stuart
Picture Researcher: Adrian Bentley
Production Manager: Matt Weyland
Indexer: Kay Ollerenshaw

Printed and bound in Hong Kong
1 2 3 4 5 01 02 03 04 05

CONTENTS

INTRODUCTION

Maps are one of the forms taken by the human impulse to record the world and relate its parts to one another. Even very early maps were conceptually sophisticated, since they were based on an imagined bird's eye view of the world which was then described in terms of conventional signs and symbols. If, as many experts believe, scratched lines on some prehistoric bones were intended as route indicators or settlement plans, mapmaking was a significant step forward in the intellectual development of human beings.

Maps have been made for many purposes, from establishing land ownership to waging propaganda wars. The beauty and fascination of many early maps stem from what they tell us about the state of human culture—and also about the extent of human error. Most cartographers have been curiously re-luctant to admit their ignorance, and during the medieval and early modern period, they commonly filled in blank areas with invented geography or colorful images of humans and animals, real and mythical. The images, often pleasing or picturesque in themselves, shed a good deal of light on the myth-making imagination; like invented geography, they reveal some of the hopes and fears of the culture that generated them.

It is of course true that maps also serve as records of advancing knowledge. From medieval times, despite occasional aberrations, coastlines were delineated with increasing accuracy and details of interiors more fully recorded. Strange races and mythical creatures disappeared from maps of land areas, although monsters continued to populate the sea for much longer, eventually as purely decorative devices. By the seventeenth century, maps were being made that combined visual splendor with an accurate coverage of many parts of the world; the lavish atlases published by the Blaeu family in Amsterdam are widely regarded as the most beautiful map collections ever produced.

In the eighteenth century, the last great myth was dispelled when Captain Cook proved that there was no great Southern Continent at the bottom of the world. The location of geographical features was now being fixed within a grid of coordinates by new instruments and ambitious national surveys. As more and more information was incorporated, images disappeared from most political and physical maps, and even thematic maps, with their standardized symbols for units of population or mineral resources or diseases, became severely, admirably functional, but of rather limited visual appeal. Fortunately, in the even more technologically advanced age of computers, satellites, and space probes, the trend has, to some extent, been reversed. Novel maps are constantly devised, and when El Niño or the surface of Mars are displayed in "false" or computer-enhanced colors, the results are as spectacular as in the great maps of the past.

Consequently, this book is illus-trated by maps from most phases of western history between 600 B.C. and A.D. 2001. Each map appears on a double-page spread along with a brief description of its significance and a visual feature picking out some of its key elements. The double page that follows is devoted to placing the map in history, the development of cartography, or world exploration. If it is successful, this arrangement will offer a wide-ranging introduction to one of the great adventures of the human mind.

BABYLONIAN WORLD MAP

The Babylonian world map is the earliest surviving map to represent more than localities. It records many of the cities and states of ancient Mesopotamia and neighboring regions, but its prime purpose seems to be to relate these to mythical lands beyond "the salt sea."

For over a millennium, Babylon was the greatest city of the ancient Near East. Even at times when it was in political eclipse, it remained an important cultural and religious center. Its importance has been confirmed by archaeological finds of thousands of clay tablets—the "paper" of Mesopotamian civilization—incised when damp with a wedge-shaped stylus to produce a type of writing known as cuneiform. The tablets include a handful of maps, among them an early (*c.* 2300 B.C.) city plan and the sixth-century B.C. Babylonian world map

illustrated here. The world is represented as a flat disk surrounded by water, a conception that was to persist for almost two thousand years into the Christian Middle Ages. The "world" is very much Mesopotamia-centered, ignoring peoples such as the Persians and Egyptians who were well known to the Babylonians. As the fragmentary text on the tablet makes clear, the preoccupations of the mapmaker were mythical rather than geopolitical, relating Babylon to a region "where the sun is not seen" and to other fabled places beyond the surrounding ocean.

LEGENDARY BEASTS
The best preserved of a series of triangular signs, arranged round the circle of "the salt sea." Each sign represents a place beyond the known world where a legendary beast was said to live.

CITY ON THE RIVER
The large oblong is marked "Babylon." The city is shown astride a double line that can only be the Euphrates River, flowing down toward the sea.

WARLIKE NEIGHBOR
The oval represents Babylon's militaristic neighbor and long-time overlord, Assyria. The circles stand for other important Mesopotamian cities.

KINGLY TRIBE
The marshlands of southern Mesopotamia are described in cuneiform as the territory of the Bit Yakin, one of the powerful Chaldean tribes from which the kings of Babylon were drawn.

THE EARLIEST MAPS

The Babylonian world map was exceptional in its scope and symbolism. Most surviving early maps represented small areas and were made for entirely practical purposes. Surprisingly, examples are known not only from early civilizations but also from prehistoric times. The ancient Greeks were the first people to adopt a scientific approach to cartography.

Above: Discovered in modern Iraq at the start of the twentieth century, the ruins of Babylon reveal a thriving city of substantial mud-brick buildings.

Mapmaking involves advanced skills and attitudes, notably the use of symbols to represent real things and the ability to visualize the world in abstracted and scaled-down form. Nevertheless, as many as fifty sets of prehistoric marks on rocks or bones have been interpreted as simple maps, though most such interpretations are controversial. There seems little doubt, however, about a wall painting from Çatal Hüyük, a settlement in Anatolia (modern-day Turkey) that prospered in the late seventh millennium B.C. by trading obsidian. One of many wall paintings on the site shows, in plan, a group of rectangles whose resemblance to the layout of the houses excavated by archaeologists seems too close to be accidental. Creating a mental map of the settlement may have been encouraged by the fact that houses at Çatal Hüyük were clustered together and were entered via their flat roofs, so their inhabitants would have regularly seen them from something like a bird's-eye view.

The inherent difficulty of mapmaking is confirmed by the existence of (at most) a handful of maps among many thousands of artifacts. This remained true of the great early civilizations of Mesopotamia and Egypt. Quantities of writing, in the form of incised clay tablets, have been recovered from the site of ancient Babylon. The tiny percentage of surviving maps are mainly simple town plans or records of land ownership, boundaries, canals, and similar features of vital importance in an economy based on irrigated agriculture. The earliest Babylonian map is believed to date from about 2300 B.C. The "world map" from 600 B.C. represents a real mental leap forward, showing—albeit in diagrammatic rather than

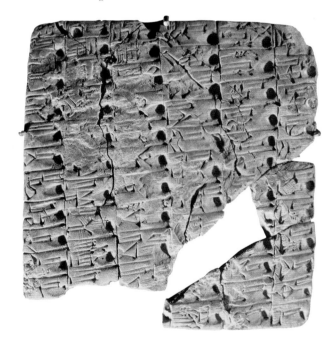

Above: This clay tablet dating from the fourth millennium B.C. shows the characteristic wedge-shaped cuneiform script used for writing in Babylon and throughout Mesopotamia.

geographically accurate form—states in relation to one another and to mountains, rivers, and marshes.

Maps were also rare in ancient Egypt, where the mindset seems to have preferred written lists to plans. Apart from items such as generalized paintings of the Nile and "maps" of the underworld through which the dead must find their way, the most interesting artifact is the Turin map, dated to about 1150 B.C. Now in two separate parts, it shows the mountains east of the Nile where gold was mined, along with the location of the miners' huts, wells, and the road network that linked the region with the Nilotic heartland of Egypt.

The Greek mindset

The Greeks were far more intellectually adventurous. This makes the absence of surviving Greek maps all the more disappointing, although convincing reconstructions have been made to illustrate how great Greek thinkers visualized the world. Like the Babylonians, Hecataeus of Miletus (*c.* 500 B.C.) held that the Earth was a flat disk surrounded by water. Although this tradition persisted into the Middle Ages, it was challenged by Herodotus, "the father of history," as early as the mid-fifth century B.C. In the first written reference to maps, Herodotus mocked the notion that the Earth must be perfectly symmetrical.

Other Greek thinkers concluded that the Earth must be a sphere, a theory fully set out by the fourth-century B.C. philosopher Aristotle. A century later, Eratosthenes calculated the Earth's circumference with only a small degree of error. Though Greek mariners rarely ventured beyond the Mediterranean, knowledge of the world gradually expanded. A Carthaginian, Hanno, was said to have sailed around Africa in about 470 B.C., and about 340 B.C. one Pytheas of Massilia (Marseille) made a voyage out into the Atlantic and beyond Britain, possibly to Iceland and almost certainly into the Baltic. More certainly, the all-conquering Alexander the Great and his army explored Asia as far as the Indus.

The accumulated cartographic achievements of the Greeks are incorporated in the works of Claudius Ptolemy (A.D. 90–168), a crucial figure in both geography and astronomy. The Greeks had evolved the idea of imposing a grid of meridians and parallels on maps, and Ptolemy listed 8,000 places with their coordinates. With the collapse of the ancient world, his work was unknown to Europeans for centuries. Its recovery was to have momentous consequences.

Above: This tile relief of a bull adorned the Ishtar Gate, the imposing ceremonial entrance to Babylon constructed by Nebuchadnezzar II. The emperor's reign, from 605 to 562 B.C., marked the height of Babylonian power and prestige.

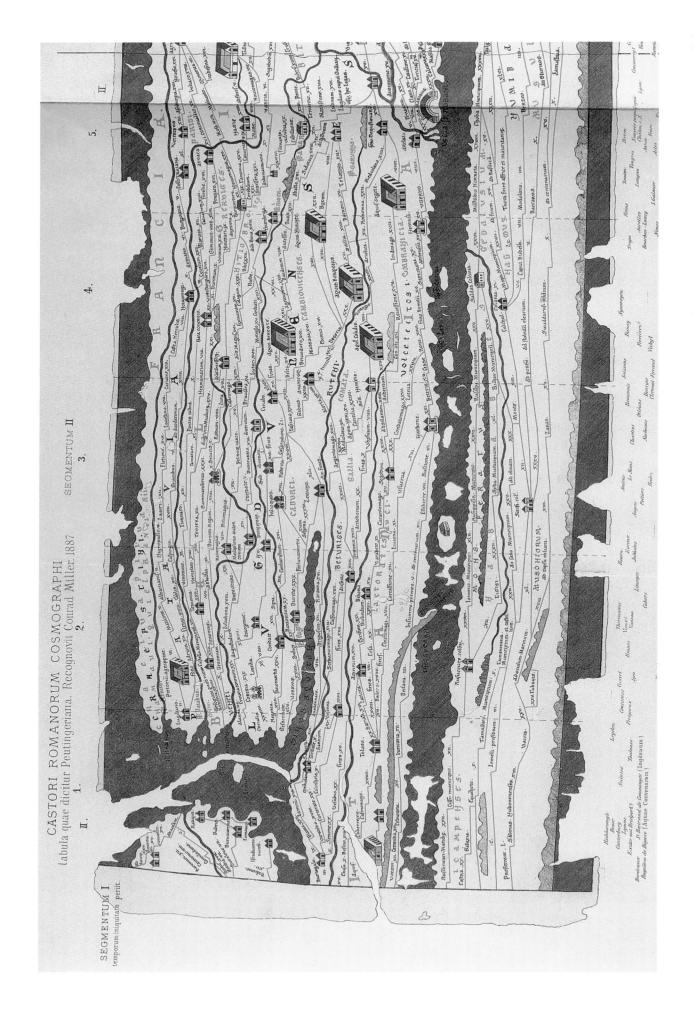

CASTORI ROMANORUM COSMOGRAPHI
labula quae dicitur Peutingeriana. Recognovit Conrad Miller. 1887.
SEGMENTUM II

THE PEUTINGER MAP

The Peutinger map provides a unique view of the ancient Romans' achievement as road-builders. It records a vast network linking towns and staging posts from Britain and Gaul to Armenia and Arabia. The map's unusual strip form reflects its function as a guide to distances and connections, comparable to modern freeway and subway plans.

Though now divided into eleven sections, the Peutinger map was originally a single roll of parchment, 22 ft. (6.7 m) long but only 13 in. (34 cm) wide. To fit most of the known world into this format, the Roman cartographer stretched and squashed the contours of countries and provinces out of all recognition, creating a narrow east–west version of the known world. In a general map this would have been hopelessly misleading, but here the intention was simply to record the existence of roads and the distances between settlements or staging posts. Some other features, such as harbors, spas, and rivers, are designated by symbols, and the primacy of the three great cities of Rome, Constantinople, and Antioch is conveyed by pictures of enthroned figures. The map is named after one of its owners, the sixteenth-century German scholar Konrad Peutinger. It is an eleventh- or twelfth-century copy of a Roman original that was made in the fourth century A.D. By the time it was transcribed, the first sheet, with most of Britain and parts of Spain and North Africa, was evidently missing. The section shown here is the first to have survived.

ROMAN TOWNS
The east and south coasts of Britain appear to run north–south. Among Roman British towns shown are Camuloduno (Colchester), Dubris (Dover), Duroaurus (Canterbury), and Iscadumnonioru (Exeter).

ANCIENT SPA
The symbols in the form of buildings around a central pool denote spas. They include Aquae Calidae, which continued to function 1,500 years after the fall of Rome under its French name, Vichy.

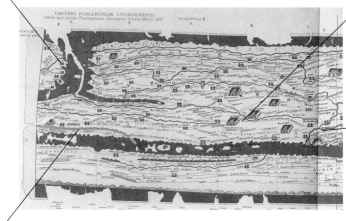

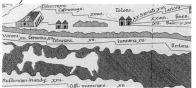

MOUNTAIN MARKERS
Mountains are marked in brown on the Peutinger map; these are the Pyrenees. Most of Spain, like Roman Britain, will have appeared on the missing first segment.

PORT SYMBOL
The arc-like symbol represents a port (Latin *ostia*), in this instance Fossae Marianae (Fos-les-Martigues) on the southern coast of France. It was linked by a canal to the Rhone estuary, clearly visible to the west.

ROMAN MAPMAKING

The ancient Romans conquered a vast empire, centered on the Mediterranean, that endured for more than 600 years. Most of their maps are lost, and what has survived is mostly filtered through medieval copyists. Even from this it is clear that officials, soldiers, and travelers were able to make use of an impressive, empire-wide communications network.

Above: Marcus Vipsanius Agrippa, shown here in a marble bust, served Augustus as a general, naval commander, and administrator; relatively humbly born, he became Augustus' son-in-law.

At some time before his death in 44 B.C., Julius Caesar ordered a survey to be made of the known world. It was completed under his successor, Augustus (emperor, 27 B.C.–A.D. 14), who delegated the task to his right-hand man, Marcus Vipsanius Agrippa. His world map, which presumably took the form of a large block of engraved marble, was set up in one of the central colonnaded areas in Rome, and it is possible that copies were also made for other imperial cities. No trace of Agrippa's map has been found, but references in Roman literature leave no doubt as to its fame and enduring influence. It may well have provided the basic information for the Peutinger map; for this, though mainly reflecting the fourth-century situation, includes some first-century elements that had become obsolete, notably the towns of Pompeii and Herculaneum, which had been obliterated by an eruption of Mount Vesuvius in A.D. 79.

Streets and properties

Some idea of the size and scope of Agrippa's map can be gained from what is known of the *Forum urbis Romae*, a third-century A.D. plan of the city of Rome engraved on marble. Substantial fragments of the plan, and the wall that held it, have survived, enabling scholars to reconstruct it in its entirety. Oriented roughly south–east, it measured over 42 ft. (13 m) high and 59 ft. (18 m) wide. It is also impressively accurate and generally drawn to scale, although important buildings tend to be made disproportionately large. Employing a range of sophisticated symbols, it was evidently a characteristic product of the Romans' practical genius.

The same is true of their cadastral (land registration) records, which were needed to define boundaries, settle disputes, and provide a basis for tax assessments. Many of these seem to have been engraved on bronze, a material that rarely survives because it is so easily melted down by later generations for reuse. Fragments of cadastral maps in stone were found at Orange in France, but the most extensive records appear in medieval copies of the *Corpus Agrimensorum*, a collection of land surveys accompanied by little picture maps and ground plans. Picture maps are also among the illustrations to another medieval manuscript, the *Notitia Dignitatum*, which lists the empire's administrative personnel around A.D. 400.

Route maps

Unlike the Greeks, with their mathematical and scientific bent, the Romans seem to have had little interest in the theory of cartography. Their itineraries and route maps are prime examples, focused on how to get from place to place, though even the Roman mapmaker yielded to the temptation to include some incidental information.

The Romans distinguished between the *itinerarium*, or written itinerary, and the *itinerarium pictum*, the painted itinerary, or road map. Written itineraries—essentially lists of places with the distances between them noted—were more common; thanks to the efficient road system, with its regular milestones, they were all that was needed for most purposes. Official itineraries were prepared by an imperial institution, the Cursus Publicus, that was responsible for travel arrangements and mail deliveries. An official document was probably the basis for the "Antonine

Above: The remains of a Roman road cut through the ruins of Carthage, in present-day Tunisia. The Romans' road network ran throughout Europe, North Africa, and the Middle East.

Left: The ruins of Pompeii. The southern Italian city appeared on the Peutinger map even though it had been destroyed by the eruption of Vesuvius some 300 years earlier.

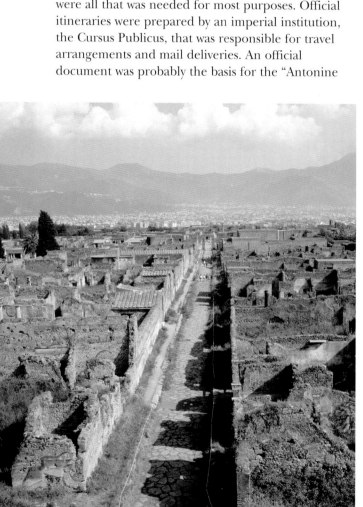

Itinerary" (late third century A.D.), in which a variety of materials has been added to an itinerary prepared for the Emperor Caracalla's journey to Egypt in 214–215. The conversion of the empire to Christianity led to the appearance of the first-known pilgrim itinerary, from Bordeaux to Jerusalem, made in A.D. 333.

Itinerary maps, as opposed to written lists, were evidently common. As late as A.D. 383, an official named Vegetius described the importance of itineraries to a military commander, especially in their visual form as an *itinerarium pictum*. The Peutinger map is the only example to have survived, so it is impossible to know whether its apparent idiosyncrasies were typical or not. The only item that resembles it is a scrap of parchment that was part of a larger skin, used as a decorative cover for a shield. Discovered at Dura-Europos in modern Syria and dating from before A.D. 260, it bore a schematized map of a stretch of the Black Sea coast, with staging posts represented and distances noted. Like other tantalizing fragments of antiquity, it leaves many questions unanswered.

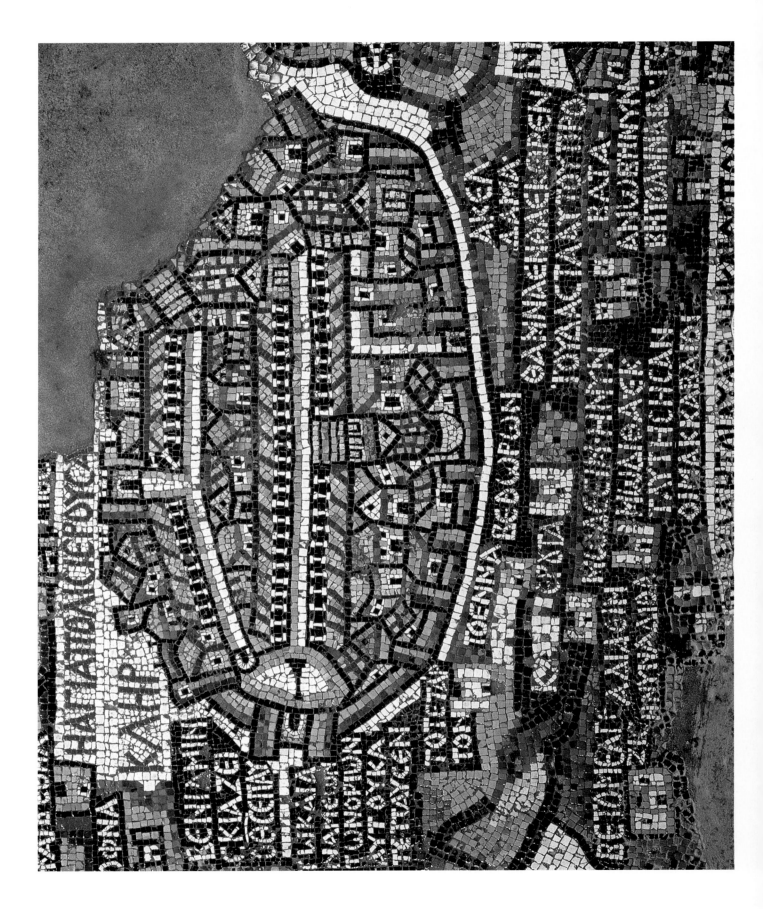

MADABA MOSAIC MAP

Found in a church at Madaba in Jordan, this sixth-century map shows the late Roman tradition being modified by Christian and Eastern influences. Boldly executed in mosaic, the surviving fragments include this striking view of the Holy City, Jerusalem.

Mosaics are made with small colored pieces of stone or glass, which are placed side by side in a concrete base to build up a picture. The Romans used the technique extensively to create durable decorative surfaces for floors and walls. The mosaic map was discovered in 1896 on the floor of a Greek Orthodox church at Madaba. It dates from A.D. 542 to 565, and its original measurements are believed to have been 20 x 79 ft. (6 x 24 m). The main surviving fragment shows Palestine from the Jordan Valley to the coast, which then extends as far as the Nile Delta. Jerusalem stands out because of its disproportionate size and the pictorial style used to represent it. It is oval-walled, with a colonnaded main street, some other important streets, and thirty-six buildings, some of them so carefully delineated that they can be identified. The city is laid out with the east at the top. The general view is from the west, but the main street is shown as though flattened, so that the colonnade on its near side appears to be upside down. The most imposing building here, the Church of the Holy Sepulchre, also appears to be upside down, so that it extends downward from its façade on the street to its dome next to the city wall.

MAIN STREET
The city's main street, running north–south. The emphasis on the roofed colonnade running down each side leaves no doubt as to its importance.

NEW CHURCH
The New Church of the Theotokos, built by the Emperor Justinian. Its consecration in A.D. 542 gives the earliest possible date for the mosaic map.

CHRISTIAN CHURCH
The Church of the Holy Sepulchre, built over the tomb of Christ by the first Christian emperor, Constantine. The building is tilted away from the main street, with the dome shown at the bottom.

CITY GATE
The north gate is depicted as the principal entrance to Jerusalem, leading to a large square and a wide north–south road. The large column that stood in the square is clearly shown.

MAPPING THE HOLY LAND

The Madaba map, Christian and Greek in character, epitomizes the changes that overtook the late Roman Empire. With the conversion of Rome to Christianity, new values emerged and Palestine (the Holy Land) acquired a previously unthought-of significance. The collapse of the Western Empire and a decline in cartographic skills served to make religious ideas and images still more influential.

Right: The Old City of Jerusalem, home of both Muslim and Christian sacred sites, still attracts hundreds of thousands of pilgrims.

Below: Emperor Constantine is baptized as a Christian in this twelfth-century re-creation of his conversion in the fourth century.

In A.D. 313 the Emperor Constantine gave Christianity an official and favored status within the Roman Empire. It soon became the dominant creed, and by the end of the fourth century, the sole-permitted one. Among the many consequences was the elevation of Palestine to a central role in human history. As the setting for the Old Testament stories and for the life and death of Jesus, the sites of the Holy Land evoked both curiosity and veneration. Pilgrimages to Jerusalem were popularized in the fourth century by the journey made by Constantine's mother, Helena, in 326. Copies survive of a written itinerary, dating to 383, that lists a series of staging posts between Bordeaux and Jerusalem.

The Western Empire collapses
In the fifth century the western half of the Roman Empire collapsed under the impact of Germanic tribal invasions. The newcomers were rapidly converted to Christianity, but both practical skills and learning declined. The eastern half of the empire survived, with its capital at Constantinople (Istanbul in modern-day Turkey).

The principal language of the eastern Mediterranean was Greek, and as time went on, the Eastern Roman Empire became increasingly Greek in character. Language, imperial, and religious customs differentiated it from the Latin West and from older Roman traditions—so much so that historians prefer to call this eastern "Roman" state by a Greek-based name, the Byzantine Empire.

The few Byzantine maps that have survived suggest that, as in the West, cartographic skills were in decline, or at any rate less valued than the religious framework in which maps were set. A sixth-century mosaic in a church at Nicopolis in Greece shows the Earth surrounded by "ocean," just as it was on the Babylonian world map more than a thousand years before. However, the Earth is represented not as a disk but as a rectangle. This is also the case in the *Christian Topography*, a book written by Cosmas Indicopleustes ("Indian sea traveler"), where the maps show a flat, rectangular world lying in a curious, box-shaped universe. Interestingly, Cosmas rails against the rival conception of the Earth as a sphere, holding that such a belief was

Above: The personification of Spring, from a mosaic found in one of the many churches in Madaba, which was an important center of Byzantine Christianity.

contrary to the cosmological information in the Bible. The dogmatic basis of Cosmas's map of the world is emphasized by the inclusion of features such as the four rivers flowing from the east out of paradise, which would appear again and again in medieval maps.

The Madaba mosaic map dates from about the same period as Cosmas's writings, but it is closer to the Roman tradition in its inclusion of some staging posts for travelers; its use of easily understood symbols for towns, churches, oases, and other features; and its relative accuracy. Nevertheless its primarily religious purpose is clear: it originally represented the regions associated with the Twelve Tribes of Israel, from Byblos and Damascus to Egypt. It is filled with biblical references, with explanatory captions in Greek; and a pictured Jerusalem is obviously intended as the main focus of the entire map.

The seventh-century Islamic conquest of the Near East sealed off the holy places from both Byzantines and western Christians. In spite of this, when the medieval civilization of the West began to produce maps again, the majority, like the Psalter map (page 28), continued to be dominated by a biblical world view.

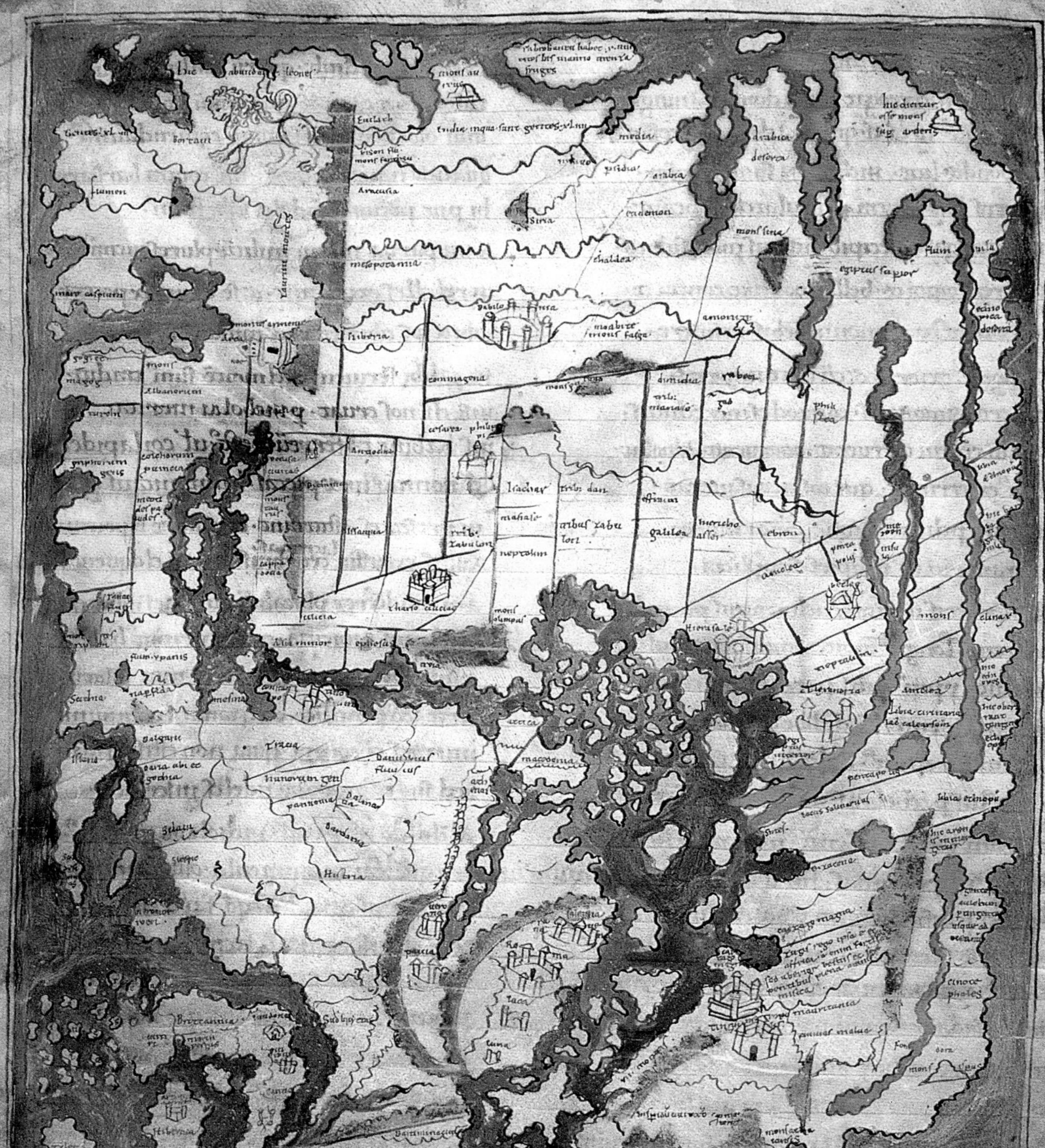

hic abundant leones
mons aureus
India in qua sunt gentes XLIIII
media
arabia
deserta
mesopotamia
chaldea
mons sina
iberia
commagena
Isachar
trib. dan
galilea
Alexandria
macedonia
pannonia
Brittannia
Hibernia
mauritania
cinocephales

THE COTTON MAP
"ANGLO-SAXON" WORLD MAP

The Cotton map presents an apparently strange world, although it becomes somewhat clearer once the viewer realizes that it is oriented so that east is at the top. Despite its shortcomings, the Cotton map was closer to geographical reality than almost all medieval maps of the world.

The map takes its nickname from the fact that it was drawn in Anglo-Saxon England. Most scholars date it to the late tenth century, although some place it a few years later. "Cotton" refers to Sir Robert Cotton (1586–1631), one of the maps early owners. Measuring 8 by 6½ in. (21 by 17 cm), it is part of a Latin geographical work, *Priscian's Periegesis*, but map and text have little apparent connection. Evidence indicates that it was drawn in England, but its ultimate source may be a lost Roman map, which would explain why it is less schematized and fable-filled than later medieval world maps. The continents are shown surrounded by water, an ancient idea that would prevail for centuries. Names of Roman provinces are used, and the straight frontier lines appear to represent Roman provincial boundaries. Some features of the Cotton map, such as the extraordinarily long-armed Black Sea, appear on later medieval maps, but in general it is an oddly isolated production.

LIONS' DEN
The inscription notes that "Here lions abound." It is the only picture of a living creature—something that was to become a notable feature of later medieval world maps.

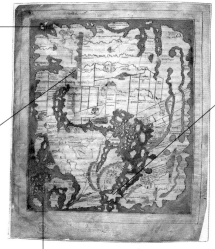

PILLARS OF HERCULES
The Pillars of Hercules was the ancient name for the promontories on either side of the Strait of Gibraltar. Here they are shown in visually literal pillar form.

NOAH'S ARK
Noah's Ark at rest upon Mt. Ararat after the Flood had subsided. On this map, the scribe has used green as the coding color for mountains.

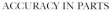
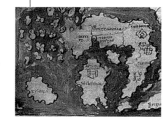

ACCURACY IN PARTS
Northern Europe, the British Isles, and Iceland are shown with an accuracy not seen again for centuries, even if Scotland and Ireland are skewed round and aligned east–west instead of north–south.

EARLY MEDIEVAL MAPS

Because it was relatively close to geographical reality, the Anglo-Saxon map is believed to have been inspired by a lost Roman original. Most other post-Roman mapping in Europe was more abstract and schematized, reflecting the values and preoccupations of the Church.

The Roman Empire in the West collapsed during the fifth century A.D., brought down, without any conscious intention, by "barbarian" peoples. Among these were the conquerors of the Romanized British—the Anglo-Saxons, by or for whom the Cotton map was made.

Folk wanderings went on long after the fall of the empire, as Franks, Vikings, Magyars, and others carved out territories for themselves, while the forces of Islam swept across North Africa and into Spain. Only in the later tenth century did Christian Europe achieve a measure of stability. By then, the mapmaking skills developed by the Romans had been largely forgotten, while the Romans' own maps had been destroyed or were falling to pieces.

Encyclopedic maps

The peoples of the successor states, with their simpler, subsistence-based way of life, had only restricted cartographic needs, at first usually satisfied by written descriptions (for example, of property boundaries).

Above: An empire in crisis: Ostrogoths led by King Vitigern attack Rome in 537 in this nineteenth-century illustration. From 376 on a series of invasions by "barbarian" peoples—Visigoths, Vandals, Ostrogoths—combined with internal difficulties to weaken and then destroy the Western Empire. Roman mapmaking skills disappeared with the civilization that had created them.

Literacy was found mainly in the church, and monastic scribes preserved substantial amounts of the thought and literature of the ancient world. Consequently, most surviving early medieval maps were attached to encyclopedic or theological works. Some of these enjoyed centuries of popularity and were copied many times.

Even the maps in the earliest manuscripts were not necessarily drawn by their authors, however, and they often vary from copy to copy. The majority were not workaday local or regional maps, but displayed the entire world in a more or less schematized form. Such maps were of two basic types. One, the zonal map, showed the world as a circle, divided into five or seven horizontal bands. These represented the frigid, temperate, and torrid zones north and south of an "equator" in the form of an ocean or river. By contrast with the north-oriented zonal map, the T-O map, as it is known today, was drawn with the east at the top. It, too, was most often circular, but occasionally it was shown as an oval or rectangle. The modern term "T-O" describes the circle or disk (O) divided into three segments by fitting a T into it so that each extremity touched the circumference. The upper segment represented Asia; the two lower segments, Europe and Africa. This kind of map could be presented as a purely geometric figure, or drawn with slightly more awareness of geographical realities, in which case the horizontal bar of the T appeared as the Don and Nile rivers, while the Mediterranean took the place of the upright bar of the T.

Isidore of Seville

The most map-illustrated of early medieval authors was a seventh-century encyclopedist, Isidore, bishop of Seville. Manuscripts of his work carry "world maps" that range from a completely schematized T-O—with nothing but the names of the continents written in the segments of a circle—to elaborate versions which attempt to delineate coastlines, islands, and mountains, and little pictures representing places, people, and creatures.

The circular maps associated with Isidore and other authors were to have a long history. More unusual were the Beatus maps, attached to manuscript copies of an eighth-century commentary on the Apocalypse of Saint John by the Spanish writer Beatus of Liebana. Some belong to the circular or oval T-O type, but most are roughly rectangular, and much more neatly and spaciously laid out than Isidorean maps. All of the rectangular examples are believed to derive from a single lost map illustrating the travels of the Apostles. Adam, Eve, and the serpent are shown in some detail at the top of the page (that is, the east, the traditional location of the Garden of Eden). A broad, red line representing the Red Sea marks off a fourth, fabled continent, the Antipodes.

Down to the twelfth century there was no clear line of development in the medieval type of world maps, though some examples of the T-O type seem to reflect a desire to include more pictorial elements and greater detail. Then in the thirteenth century a group of fascinating maps appeared—among them the Psalter map and the Hereford *mappamundi* (see pages 28 and 40)—that brought the medieval tradition to its culmination.

Above: Scribes, usually monks, were the chief guardians of ancient and medieval learning. This cathedral scribe was carved in Mantua, Italy, in the fifteenth century,

WORLD MAP
AL-IDRISI

While mapmakers in Christendom were bound by theological and didactic concerns
that made geography a secondary consideration, Islamic cartographers remained
in touch with the ancient Greek tradition. Al-Idrisi was a particularly distinguished
figure, working at a meeting place of two cultures.

Sicily was conquered in 827 by the Arabs, who held it until it was taken in 1072–1091 by Norman adventurers. The Arab presence on the island remained strong, and the Norman policy of toleration turned Sicily into a prosperous society at the crossroads of Mediterranean commerce. It was also a center of scholarship, especially under King Roger II (1097–1154). One of those invited to Roger's court at Palermo was al-Idrisi, who had studied at the University of Córdoba in Muslim Spain, traveled in

Central Asia, and, most unusually, visited England and France. For fifteen years al-Idrisi labored to create an account of the world. In 1154 he completed *The Book of Roger*, manuscripts of which were written in Latin and Arabic, along with a 73-sheet world map. Engraved on silver, the map was regarded as a wonder until it was destroyed in a riot in 1160. The small, relatively unambitious south-oriented map reproduced here also accompanied *The Book of Roger* and gives some idea of the quality of al-Idrisi's work.

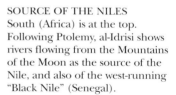

SOURCE OF THE NILES
South (Africa) is at the top. Following Ptolemy, al-Idrisi shows rivers flowing from the Mountains of the Moon as the source of the Nile, and also of the west-running "Black Nile" (Senegal).

ACCURACY
The Nile Delta and Arabia are convincingly represented, like most of the Mediterranean, the Black Sea, and even the Caspian. Surprisingly, since Arab ships sailed the Indian Ocean, al-Idrisi shows the region only perfunctorily.

LOST ATLANTIC
Islands, but no British isles: the Atlantic lay on the edge of al-Idrisi's world, and the cartographer's visits to England and France evidently failed to make any great impression on his mental map.

CLIMATE ZONES
The parallels on al-Idrisi's map (representing climate zones rather than latitude) were the logical consequence of his conviction that the Earth was a sphere, though here he represents it in conventional form as a disk.

THE RISE OF ISLAM

Al-Idrisi was one of the men of genius fostered by a civilization larger and more brilliant than that of medieval Europe. From the seventh century, when Muslim armies burst out of Arabia and began to overthrow long-established states, Islam was a potent force whose sway eventually extended from Spain to India and beyond.

The doctrines of Islam were transmitted to the world by the prophet Muhammad, who began his mission in the Arabian city of Mecca. Local hostility prompted his flight to neighboring Medina in A.D. 622, later adopted as the first year of the Islamic calendar. A series of military campaigns led to the submission of Mecca, and by the time of Muhammad's death in 632, most of Arabia had been converted.

Fired by their beliefs and united for the first time, the Arab tribes became an almost irresistible force. They swept out of the peninsula, drove the Byzantines from Egypt and Syria, and by the mid-seventh century had destroyed the Persian Empire—something the Roman legions had never managed to do.

Islam divided

After Muhammad's death the Arabs were led by a succession of caliphs. Setting up a new society in the wake of victory increased tribal and other divisions, which came to a head after the murder in 661 of the fifth caliph, Muhammad's cousin and son-in-law, Ali. Ali's followers refused to accept the next caliph, Mu'awiya, and declared a *sh'ia* (separation). Thus began what proved to be an enduring division of the Islamic world into Sunni and Sh'ia sects.

Mu'awiya suppressed all opposition and turned the caliphate into something more closely resembling a monarchy, with its capital at Damascus. Under the dynasty he founded—the Ommayads—the tide of Arab conquest rolled on, reaching Central Asia and Spain.

The vast Muslim world prospered under the Ommayads. More willing to learn from others than most empire-builders, the Arabs took over or adapted institutions, techniques, and ideas from the Byzantines and Persians. Non-Muslims were taxed rather than

Above: The twelfth-century philosopher Ibn-Rushd, as imagined in a later painting, was one of the most influential Islamic scholars in the West, where he was known as Averroës.

persecuted. Large numbers were converted, and the obligation on all believers to learn Arabic led to the creation of a unified culture in which the distinctions between Arabs of Arabia and other Muslims were eroded. "Arabs" became, at least in the Middle East, not Arabians but people who spoke Arabic.

This trend became more pronounced after 750, when the Ommayads were overthrown and a new Abbassid dynasty ruled from Baghdad. An essential cultural unity persisted even after Islam began to fragment in the tenth century, with the emergence of a separate Fatimid caliphate in North Africa and the reappearance of a more or less independent Persia.

Throughout this period Islamic civilization was more advanced than that of medieval Europe. Muslim scholars translated the works of ancient Greek philosophers and medical authorities, and took over Indian concepts of mathematics, including "Arabic" numerals. Building on these intellectual foundations, they produced outstanding thinkers such as the philosophers ibn-Sina (known in the West as Avicenna) and ibn-Rushd (Averroës), the mathematician al-Khwarismi, and the Persian poet-

Above: This schematic plan, created on a sixteenth-century ceramic tile, shows the location of the tomb of Muhammad in Mecca, the holiest of Muslim cities.

astronomer Omar Khayyam. Geographers based their view of the world on Ptolemy, and their mapping culminated in al-Idrisi's achievements. Poetry, travel writing, and other forms of literature flourished. So did architecture, with its distinctive dome and courtyard forms. In the arts the Muslim preference for the nonrepresentational—reinforced by religious prohibitions against representing reality—led to richly decorated ceramics and metalwork, often exploiting the ornamental potential of Arabic calligraphy.

Mongols and Turks
In the eleventh century the eastern regions of Islam were taken over by a people from Central Asia, the Seljuk Turks. They ultimately repelled the Crusades launched by the Christian West but, divided among themselves, fell to devastating invasions by the Mongols in the mid-thirteenth century. This proved to be a watershed in Islamic history, paving the way for the triumph of another Turkish people, the Ottomans, whose empire would have a different character.

Above: The flight of Muhammad from Mecca in 622, an event adopted as the starting point of the Islamic calendar.

PSALTER MAP

The Psalter Map is a mid–thirteenth-century English example of a *mappamundi*—a medieval world map combining elements of geographical knowledge with religious symbolism and a variety of legendary features. A delicate painted miniature, the Psalter Map was, appropriately, part of a devotional work.

A psalter is a book containing psalms. Intended for individual use, it is usually small, and in the example here, the page is almost completely filled by a map measuring only 3¾ in. (9.5 cm) across. The map can be regarded as a preface to the psalms, picturing the Earth as part of a divinely ordained scheme of things. Christ presides over it, Jerusalem sits at its center, and the east—said to be the location of Eden—is placed at the top. Following a tradition that went back to the Babylonians, the Earth is shown as a disk. Some older cartographic conventions persist in the use of a few symbols, such as triangles to denote towns and villages. The map's relationship with reality is tenuous, however, and even the geography of Europe is correct only in a diagrammatic fashion. The fantasy elements culminate on the right-hand (African) side of the map, which carries a series of portraits of part-human races whose supposed existence fascinated Europeans for centuries. Any spaces around the map are filled with pleasant decorative touches, in particular two homely dragons with luxuriantly blossoming tails.

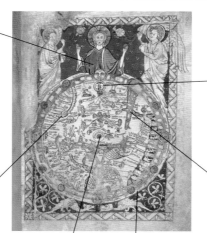
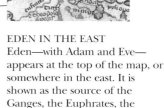

CHRIST PRESIDES
Jesus Christ, orb in hand and flanked by angels swinging censers, is a majestic figure. His presence reinforces the symbolic function of the map.

EDEN IN THE EAST
Eden—with Adam and Eve— appears at the top of the map, or somewhere in the east. It is shown as the source of the Ganges, the Euphrates, the Tigris, and other rivers.

GOG AND MAGOG
The walled kingdom of Magog appears on the left. According to legend, as the Day of Judgment approached, Gog and his followers would break through the wall and create havoc.

THE EARTHLY CENTER
Jerusalem is placed like a bull's-eye at the center of a circular Earth. This is not so much a geographical statement as a proclamation of the city's significance for Christians.

IMAGINARY BEINGS
At the lower right is shown a parade of imaginary races conceived by the medieval mind and believed to live in yet-undiscovered regions. Prominent among them are headless men with faces in their chests.

THE RED SEA
Following an already well-established medieval tradition, the Red Sea is shown in red, although other seas on the Psalter Map are depicted in green and rivers are blue.

MEDIEVAL WORLD MAPS

The Psalter Map is a miniature example of a distinctive medieval genre, the *mappamundi*. This was a world map, based on the traditional model of a disk surrounded by water, but with the details modified to reflect Christian teaching and contemporary lore. Very few maps have survived from the Middle Ages.

The *mappamundi* (plural *mappaemundi*) derives from the ancient picture of the Earth as a flat disk or "O." With east at the top, this was schematized by rendering the Don and the Nile as a straight line, forming the horizontal bar of a "T," while the Mediterranean became its upright. Such "T-O" maps allowed their makers to divide a circle into three sections that could be labeled with the names of the three known continents—Asia, Europe, and Africa. The great majority of surviving world maps are diagrams drawn in this tripartite and three-label form, or some variation on it, and they often served as simple book illustrations.

The name *mappaemundi* is often reserved for the much more elaborate examples found from the thirteenth century onward. These showed the contours of landmasses (albeit with wild inaccuracy) and filled the spaces inside them with an extraordinary mixture of contemporary and ancient place names; picture-symbols of locations drawn from the Bible, classical mythology, and popular legend; and lively sketches of half-human races and fabulous beasts. The imaginary beings were popularized by medieval books of tall tales, such as Sir John Mandeville's *Voyages and Travels*.

Maps to entertain and uplift

The overwhelming majority of maps that have survived from the Middle Ages are world maps, intended for edification rather than practical use. Their end product might be regarded as a mixture of entertainment and uplift. The element of religious instruction was always strong, though the elements included varied from map to map. Among those present in the Psalter Map are the authority figure of Jesus presiding over it, the location of Paradise, or

Eden, in the Far East at the top, and the emphasis on sites and events in the Holy Land. Sometimes, as on the Psalter Map and the Hereford *mappamundi* (page 40), Jerusalem is depicted as the center of the world.

Although the Psalter Map is small—the size of a book—many *mappaemundi* were large enough to put on display for the instruction of the faithful. The English word "map" derives from the Latin *mappa*, which actually means "cloth;" the *mappamundi*, or "cloth of the world," was evidently thought of as primarily a wall hanging.

Most surviving *mappaemundi* are made of parchment. Doubtless the large cloth versions were subject to more wear and tear and were discarded when fashions in pageantry—or knowledge of the world—changed. There are thirteenth-century references to displays of world maps at a number of places in England, and since these included Westminster Palace, Winchester Castle, and Waltham Abbey, the large cloth versions were certainly prestigious items in their time.

The English seem to have been particularly interested in *mappaemundi*, although this had no discernible effect on the accuracy with which the British Isles were depicted. The most important group of world maps made before the fifteenth century all have English connections. These include the tenth- or eleventh-century Cotton ("Anglo-Saxon") map, which may owe something to ancient Roman geography; the Psalter Map; the 1280s Hereford map; the early or mid–thirteenth-century Ebstorf map, which, though certainly drawn in Germany, is closely related to others in the group and may have been commissioned by an English patron; and the mid–fourteenth-century world maps accompanying Ranulf Higden's *Polychronicon*.

The largest and perhaps the most splendid of all known *mappaemundi* was the Ebstorf map, at more than 11 ft. (3.5 m) across. It was destroyed in an air raid in 1943, although its appearance has since been reconstructed. The largest original still in existence is the fascinating Hereford map.

Right: Waltham Abbey, burial place of King Harold I, who built it in the eleventh century, possessed its own *mappamundi*, an object of great prestige for any medieval church or castle.

MEDIEVAL ROUTE MAP

Medieval travelers had little to help them find their way from place to place. Some carried written directions, but the thirteenth-century chronicler Matthew Paris displayed a visual imagination rare for the period when he created an attractive sketch-map itinerary for those en route for Italy, perhaps on their way to the Holy Land.

Matthew Paris was a monk at the Benedictine abbey of St. Alban's, which he entered in 1217; he was probably born around 1200. He became the abbey's chronicler about 1240, writing lives of its abbots as well as his great *Chronica Majora*, which described events between 1235 and 1259, the year of his death. His artistic gift led him to decorate his works with delightful marginal drawings, and a similar impulse evidently lay behind his maps, as important in their own way as his famous chronicle. Paris left England only once, in 1248–1249, when he undertook a papal mission to reform a Benedictine house in Norway, but he was in touch with the great world through eminent visitors to St. Alban's Abbey, a famous foundation. Several versions of maps of Britain and Palestine, drawn in his own hand, survive. The itinerary map directs the traveler from London to Apulia in Italy. Each of its five manuscript pages has two columns, to be read from top to bottom. The first, illustrated here, shows London to Dover in the left column and, in the right, alternative routes from Calais to Reims and Boulogne to Beauvais.

KEY FORTRESS
Dover Castle, the mighty fortress built in the 1180s by Henry II, is already being described here as the key to England. Immediately above it, the English Channel awaits the traveler.

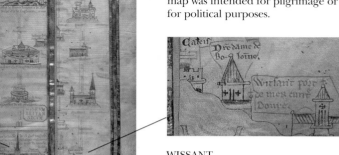

ALTERNATIVE ROUTES
Reims and Beauvais, stages on the alternative routes across France. Opinions differ as to whether Paris's map was intended for pilgrimage or for political purposes.

CAPITAL CITY
London, "the principal city of England," is the starting point of the journey. Paris's sketch shows the walls surrounding the Tower and Old St. Paul's Cathedral, with the River Thames in the background.

WISSANT
Despite its placing to the right of the column, the port of Wissant is actually usefully situated between Calais and Boulogne, the starting points for the alternative routes proposed by Paris.

EARLY PILGRIM ITINERARIES

Matthew Paris's itineraries and maps were clearly drawn with the needs of pilgrims in mind. Making pilgrimages was one of the most important reasons—or excuses—for travel during the Middle Ages, giving rise to itineraries, guides, and a limited number of maps.

V isits to holy places have featured in a number of religions, but the pilgrim was especially prominent in Christian Europe during the Middle Ages. Pilgrimages were made as expressions of personal piety, in the hope of being healed, in fulfillment of a vow, or to work out a penance. Some pilgrims may have been attracted by the possibilities of adventure and diversion, motives unacknowledged in a society where travel was expected to have a practical purpose. The pleasurable, sociable aspects of travel are certainly implicit in Geoffrey Chaucer's famous fourteenth-century work *The Canterbury Tales*, in which pilgrims swap stories on their way from Southwark in

London to the shrine of St. Thomas à Becket at Canterbury. They have the chance to tell long and lively tales: The journey on horseback is time-consuming despite the short distances involved.

However, pilgrims often traveled much farther afield. The importance of Jerusalem and the holy places there is indicated by the city's disproportionate size on the Madaba mosaic map (page 16). Pilgrims flocked there after the city was visited by Emperor Constantine's mother, Helena, in A.D. 326. After the collapse of the Roman Empire in the west and the Islamic conquest of the Near East, Rome replaced Jerusalem as a European place of pilgrimage, and a surprising number of wealthy Anglo-Saxons managed to reach the Eternal City.

Rare early records

Archbishop Sigeric of Canterbury went to Rome in 990 to receive his pallium (robe of office) from the pope. One rare surviving record is a list he made of churches he visited and places through which he passed on his way home. No map of the city is known before the twelfth century, when simple diagrams began to be drawn showing Rome's ancient walls and famous monuments.

Above: Pilgrims travel to Canterbury in this thirteenth-century stained-glass window to pray at the shrine of Thomas à Becket, murdered by followers of the English king Henry II in 1170.

Left: Canterbury Cathedral in Kent was one of the chief destinations for English pilgrims—but they often traveled much farther afield to Europe or the Middle East.

Below right: The pilgrims' destination: the cathedral at Santiago de Compostela in northwestern Spain. Built in the ninth century over a tomb claimed to be that of St. James, the cathedral has been a site of pilgrimage for more than a millennium. It continues to attract many thousands of worshipers.

and large numbers of travelers rose to the challenge. Europeans gained wider knowledge and experience as a result of the Crusades, including an ability to chart the Mediterranean coastline (Carte Pisane, page 36). Maps of Jerusalem and the Holy Land were frequently made. The medieval preference for a neat diagram rather than an irregular reality persisted, however, and almost all views of Jerusalem, for example, showed the city as a group of pictured monuments set within perfectly circular walls.

Matthew Paris's two maps of the Holy Land, like his itinerary, are of exceptional interest. One mirrors contemporary reality by giving greatest prominence to Acre, by then the last remaining Crusader stronghold. The ultimate failure of the Crusaders did not end pilgrimages to the Holy Land, and it is characteristic of the slowly widening European horizons that the first usefully detailed map of Palestine emerged from the early fourteenth-century workshop of Pietro Vesconte, which was also involved in the progressively more accurate charting of the Mediterranean.

Perhaps the most popular of all medieval shrines was that of St. James the Apostle at Santiago de Compostela in northwest Spain. Pilgrim routes to the shrine ran through France from Britain, Germany, and Scandinavia, and the way these routes functioned makes it easier to understand why so little reliance had to be placed on maps. The routes were few but known, marked by many famous Romanesque churches (Vézelay, Moissac, Autun) that owed their grandeur to the pilgrimage route, and were serviced by hospitals (offering lodging as well as medical treatment) and commercial hostelries.

Whatever the hardships involved, finding the way from place to place can only have involved relatively straightforward verbal instructions, or soon-discarded written notes, obtained at each staging post. However, one substantial manuscript work, the *Liber Sancti Jacobi,* survives from the twelfth century. It contains a section that can only be described as a guide book, noting itineraries and distances between resting places, as well as giving information about sights to be seen, currency, accommodations, and other practical matters.

A pilgrimage to the Holy Land was a more daunting prospect, but the journey became possible after the capture of Jerusalem by Crusaders in 1099,

THE CARTE PISANE
EARLY MARINER'S CHART

Medieval seafaring was a dangerous and difficult business, even in the tideless waters of the Mediterranean. Navigators relied largely on memory or written directions until the late thirteenth century. Then, quite suddenly, charts of remarkable accuracy appeared, transforming the prospects for maritime travel and trade.

In 1270, when the king of France, Louis IX, was crossing the Mediterranean, a violent storm blew up. The ship's captain consulted a chart and persuaded the king to seek shelter in a Sardinian harbor. This is the first known reference to a chart being used on board ship, and it is close to the date of the first surviving example: the Carte Pisane. The Carte Pisane and its successors are known as portolan charts—to the annoyance of cartographers, who

discern no link between sea charts and the *portolani*, or written lists of ports, distances, and directions that medieval sailors had long used. A damaged piece of parchment measuring about 20 by 40 in. (50 by 100 cm), the Carte Pisane exemplifies chartmakers' seamanlike preoccupation with coastlines and their indifference to the hinterland. Like later portolan charts, it covers the surface with rhumb lines indicating compass and wind directions.

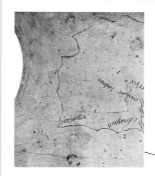

EDGE OF EUROPE
Britain is represented as a misshapen rectangle, with London in the center of the south coast. Ireland is entirely absent, and northern Europe peters out. Clearly, the chartmaker was seriously interested only in the Mediterranean.

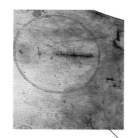

PRACTICAL INFORMATION
The inclusion of a scale bar is an indication of the practical importance of the chart. This and a second scale bar, toward the top, are placed in a circle, as was customary on portolan charts. No numerical key is given, but each subdivision seems to correspond to 10 miglia, about 7½ miles (12 km).

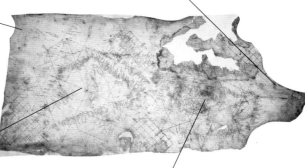

ISLAND DETAIL
The Sardinian coastline, according to one authority, is as accurate here as on a modern map. On portolan charts, coastal names are tucked away inside the coastal outline wherever possible.

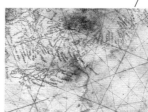

RHUMB LINES
Sixteen rhumb lines, each indicating a compass direction, fan out from a point in the Aegean Sea. There is another set close to Sardinia; the two sets are arranged to cover all the important areas of the map.

EXPANDING HORIZONS

Portolan charts such as the Carte Pisane
made navigation in the Mediterranean
safer and more efficient. Their
appearance was one of several positive
developments in seafaring that would
soon send mariners out into the Atlantic
to make the first new discoveries beyond
Europe since antiquity.

During the Dark Ages that followed the collapse
of the Roman Empire, large-scale migrations
and invasions kept Europe in turmoil. The
Arab conquest of the Near East and North Africa
disrupted shipping and trade in the Mediterranean,
and in much of Europe people reverted to subsistence
farming and towns declined in importance.

A more stable order began to emerge about A.D.
1000. The predatory Vikings and Magyars were
converted to Christianity and tamed, and though the

Muslim Arab states remained formidable, their
impulse to expand had slackened. Soon Europe
(more precisely, Latin Christendom) began to expand:
externally to the north and east; internally through
population growth. More land was settled and taken
into cultivation, and there was a revival of town life
and learning.

The expansive new spirit also manifested itself in
the First Crusade, an assault by motley European
armies on the Arab-held Holy Land. Inspired by Pope
Urban II in 1095, the Crusaders traveled distances that
were astonishing by eleventh-century standards. In 1099
they captured Jerusalem and set up "Latin" kingdoms.
These eventually fell to Arab counterattacks, but
crusades continued for another two centuries, directed
at Egypt, Tunis, and even Christian Byzantium.

Contact with the East

None of the Crusaders' objectives was achieved, but
contact with the East brought new products and new
ideas to Europe. Not least, eastern contact had a
considerable impact on seafarers. Carrying men,
horses, arms, and supplies on an unprecedented scale
stimulated seaborne traffic and improvements in ship
design. The mariners of southern Europe, accustomed
to the tideless Mediterranean and intimidated by the
strong current that sealed them off from the Atlantic
Ocean, made contact with men who sailed sturdy but
clumsy craft in northern waters. Each learned from
the other, and vessels capable of surviving on the open
ocean began to evolve.

Concurrently, navigation improved dramatically.
The magnetic compass came into use during the
twelfth century, and by 1218 it was being described as
indispensable for travel. Within a few years
Mediterranean coastlines were being recorded with
relative accuracy on portolan charts. Quite how the
information they contained was gathered remains
obscure, but within a short time schools of skilled
chartmakers were flourishing at Venice, Genoa, Pisa,
Barcelona, and Majorca.

Left: Crusaders guard Jerusalem in this illustration from the
fourteenth century. In reality the city was again in Muslim
hands, having fallen to Saladin's Mamluk army in 1247.

The most momentous outcome of these advances was that in the thirteenth century Mediterranean mariners began to venture beyond the Pillars of Hercules (Strait of Gibraltar). This led to an increase in trade with the north, and reciprocal influences led to further improvements in ship design. Of even greater importance, Mediterranean mariners for the first time sailed south into the unknown Atlantic. The earliest recorded expedition was undertaken by the Vivaldo brothers, who set out in 1291 from Genoa, apparently determined to sail right around Africa; but they were never heard of again.

Whether the earliest discoveries were made by similar expeditions or accidentally, by ships blown off course, is not clear. From the early fourteenth century, however, the islands off the west coast of Africa—the Canaries and the Madeiras—were located and visited with increasing frequency. The Azores, far out in the ocean, were also discovered, probably by wind-blown ships on their way back and forth from the more southerly islands. The discoveries were gradual and poorly documented, initially involving mainly Genoan and Majorcan seafarers. Later, Spanish, French, and Portuguese adventurers descended on the islands, hoping to found kingdoms or find gold or slaves.

By about 1380 this great triangular area of the Atlantic was well known to mariners. The breakout into the Atlantic would prove decisive for Europe's discovery of the East and West in the fifteenth century.

Below: French and English Crusaders prepare to do battle, vainly hoping to stem the Muslim advance that culminated in 1291 with the fall of Acre, the last Christian stronghold in the Holy Land.

HEREFORD *MAPPAMUNDI*

The Hereford *mappamundi* is the largest surviving world map of the medieval type, primarily symbolic and didactic in function. It is a work of considerable artistry, made still more appealing by a written "message from the author." For 700 years it has hung in Hereford Cathedral, a witness to the faith and fantasies of its time.

The Hereford *mappamundi*, or world map, was the work of a cleric, Richard of Haldingham, or Holdingham, who is believed to have drawn it in Lincoln in England at some time in the 1280s. He subsequently moved to Stoke Talmage in Oxfordshire and then on to Hereford in 1305. The map has been kept in Hereford Cathedral ever since. In the bottom left-hand corner, close to his rendition of the British Isles, Richard has used a small space to assert his authorship of this *estoire* ("history") and ask for spiritual support from his audience: "Let all who have this *estoire*, or shall hear or see or read it, pray to Jesus in God to have pity on Richard of Haldingham and Lafford, who has made and drawn it, so that joy in Heaven may be granted him." Richard was justifiably proud of his achievement, for the Hereford *mappamundi* contains such a wealth of biblical, mythical, legendary, geographical, and other material that it could be described as an encyclopedia of medieval belief compressed into a single picture.

BABYLON THE EVIL
Babylon's size evidently reflects its role in the Old Testament as the site of the Tower of Babel and an evil imperial city. Nearby, Abraham is framed by his original home, Ur.

ALL KINDS OF BEASTS
Three mythical beasts and one real one are depicted. From top to bottom they are: a centaur, sphinx, rhinoceros, and unicorn. Appropriately, in view of the sphinx's presence, Cairo is nearby on an island in the Nile.

ISLANDS IN THE MEDITERRANEAN
Mediterranean islands are given great prominence, from the Pillars of Hercules (Gibraltar) at the bottom to Delos (next to the mermaid) and Rhodes, whose long-crumbled Colossus is pictured. Note, too, Crete's famous mythical labyrinth.

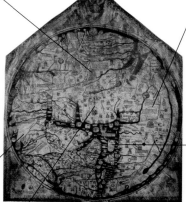

NORWEGIAN SKIER
An unexpectedly accurate sketch of a skier appears in the Norwegian peninsula, though he confronts an apparently simian creature. A bear lingers nearby, and the headland above is occupied by dog-headed men.

BIBLICAL SITES
At the heart of the map stands Jerusalem with, immediately above it, Christ crucified at Calvary. Bethlehem, the Mount of Olives, Beersheba, Jericho, and many other biblical sites are represented.

MAPPING THE MEDIEVAL MIND

With its hundreds of pictures and captions, Richard of Haldingham's map crams an extraordinary amount of information, visual and verbal, true and imagined, into less than 21 sq. ft. (2 sq. m). Its apparent confusion conceals an ordering of the most important elements that was carefully thought out.

The map is made from vellum (calfskin), a more luxurious material than parchment. It is attached to an oak frame and, including its border, measures 65 by 53 in. (165 by 135 cm). Its coloring is equally impressive, with black, red, and blue drawing inks enlivened by touches of gold leaf.

At the very top, outside the map, Christ is enthroned in majesty on the Day of Judgment. On his right, the blessed are assembling, ready to ascend. On his other side, the damned, roped together, are being led away by devils toward the mouth of Hell. All of this happens outside time and space, but immediately below, a roundel just within the map represents the Garden of Eden, with a drawing of Adam, Eve, and the serpent. In traditional fashion it is located somewhere in the Far East, beyond China and India. Close by, on the Asian mainland, Adam and Eve are shown being expelled from the garden by an angel.

The eastern orientation of the map makes it confusing at first—especially since Europe has wrongly been labeled as "Africa" and Africa as "Europe." In fact, Asia is at the top of the map, Europe at the lower

Above: The famous abbey of Mont Saint-Michel was built in 966 on a spectacular island rock just off the coast of Normandy. One of several places of pilgrimage named in Richard's map, it was also a fortress, often besieged in vain by the English.

Left: Putting Paradise on the map: the Garden of Eden, shown here in a seventeenth-century German engraving, appears on Haldingham's map with other biblical and mythical sites, distributed fairly evenly among actual contemporary places.

left, and Africa lower right. Apart from the vividly colored Red Sea and Persian Gulf, the principal seas are shown as a lop-sided T-shape, the original blue faded to sepia, in the bottom half of the map. The largest, upright stretch of water is the Mediterranean, with the Adriatic shown as an inlet next to a far-from-boot-shaped Italy. Of the three vertical arms above the Mediterranean, the left-hand one is the Black Sea (with the mythical Golden Fleece spread out beside it at the far end), the Aegean lies to its right, and the third arm is a (nonexistent) gulf in the coastline between the Levant and Egypt.

Up-to-date details

Although Richard's map contains many errors and fantasies, he evidently made an effort to include up-to-date information, especially where it was relevant to his own calling. Edinburgh may be placed on the wrong side of a near-island Scotland, but a surprisingly large number of places in Western Europe are named, notably shrines such as Mont St.-Michel in France and Santiago de Compostela in Spain, as well as the towns situated on the pilgrim routes that led to them.

There also seems to be a fundamental plan underlying the wealth of detail. Five key items—the Garden of Eden; Babylon; Jerusalem and Calvary; Rome; and the Pillars of Hercules (Gibraltar)—are placed at roughly equal distances from one another, running directly down the center of the map. They evidently form a kind of implicit narrative, tracing humankind from its beginnings, through key episodes in the Old and New Testaments and the triumph of the Church, to the edge of the world, beyond which lies only the impassable ocean.

As if to give authority to his picture of the Earth, Richard has surrounded his personal statement in the bottom left-hand corner with a drawing of the Roman emperor Caesar Augustus. He is shown as a pontiff-like figure handing three surveyors a document with an official seal. It instructs them to go out into every continent and report back on what they discover. Captions on the borders of the map elaborate on their supposed achievements, and it is hard to escape the conclusion that Richard intended viewers of his map to believe that it represented a faithful summary of the surveyors' findings, with the stamp of authority.

WORLD MAP
RANULF HIGDEN

Mappaemundi, diagrammatic world maps mixing myth and religion with some geographical information, went on being made all through the later Middle Ages. They continued to be particularly congenial to English chroniclers and to the scribes who made copies of their works.

Ranulf Higden was a fourteenth-century Benedictine monk from Chester in England. His world history, the *Polychronicon*, was highly popular, and about 120 manuscript copies of it have survived. They are illustrated by *mappaemundi* of various shapes, including the oval version reproduced here, which dates from the late fourteenth century. It is believed to be the closest Higden map to the lost prototype. With its wavy outlines and bulging shapes, it is, if anything, even more schematic than some

earlier world maps such as the Hereford *mappamundi*. Any geographical knowledge that may have reached England—via the Crusades, the development of Mediterranean chartmaking, or Atlantic island discoveries—finds no echo here. Unlike Higden, most later European examples of *mappaemundi* were influenced by portolan charts: mapmakers still presented a schematic account of the three known continents but incorporated a relatively accurate outline of the Mediterranean and the Black Sea.

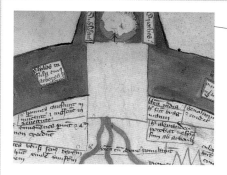

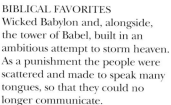

VACANT EDEN
For some reason the map was not finished. Although the panel representing the Garden of Eden looks empty, there are faint traces of a preparatory sketch showing the traditional scene with Adam, Eve, and the tree with the fatal fruit.

BIBLICAL FAVORITES
Wicked Babylon and, alongside, the tower of Babel, built in an ambitious attempt to storm heaven. As a punishment the people were scattered and made to speak many tongues, so that they could no longer communicate.

ENGLISH TOWNS
The English background of the map is apparent in the size and vivid redness of the country, and in the disproportionate number of town symbols in it.

WIND AND WATER
One of twelve wind heads round the edge of the map. The west wind blows vigorously toward the Straits of Gibraltar, which are pictured literally as Pillars of Hercules.

MONASTIC BOOKS AND MAPS

The 120 versions of Ranulf Higden's chronicle were made by many hands—mostly by monastic scribes who copied the text and drew versions of its map. For a thousand years monks performed a vital task in preserving old and creating new texts, and their labors were widely disseminated by copyists.

A publishing industry grew up in ancient Rome, where teams of well-educated (usually Greek) slaves were employed in producing multiple copies of books. Speed was important, since the absence of copyright laws meant that if a rival could obtain the text, there was nothing to stop him putting it on sale first. Haste evidently increased the number of mistakes made in copying—at any rate, authors were loud in their complaints.

In ancient Egypt, Greece, and Rome, books were written on papyrus, a paper-like material made from dampened and pounded Nilotic reeds. The books themselves consisted of long scrolls held in both hands; the right hand revealed fresh text while the left

Below Scribes in scriptoriums sometimes wrote little notes, called colophons, on their work. One complained that "Writing bows one's back and thrusts the ribs into one's stomach."

rolled up the areas that had been read. A long book took up many scrolls.

In the early fourth century A.D., Christianity became the paramount religion of the Roman world, and this coincided with a change in bookmaking. The scroll was replaced by the codex—essentially the book as it has been made ever since, with separate pages fixed together at the spine and bound with covers of wood (or, later, less heavy materials). At about the same time, papyrus was replaced by a more durable material, parchment (scraped and polished animal skins) or vellum (calf or other high-quality skins).

After the fall of Rome

In the fifth century the Roman Empire in the west collapsed. The cultural and material life of the "barbarian" kingdoms that replaced it was generally on a lower level, and the very survival of ancient literature, learning, and thought were threatened. As Europe slowly emerged into the Middle Ages, literacy was mainly confined to the Church. Churchmen often held high secular office, and most records were kept by clerks—that is, clerics.

Paradoxically, a central role in European culture was played by monks who had committed themselves to withdraw from the world. In the West monasticism effectively began with Saint Benedict, who in 528

founded the house at Monte Cassino from which the Benedictine order sprang. Since Benedict's rule insisted on the importance of work ("to work is to pray"), monasteries became centers of economic and intellectual energy.

The making of books became a monastic industry. The monks worked in a shared room, the *scriptorium*, or sometimes in individual cells or niches arranged round the cloisters. They copied gospels, missals, and similar works for use in churches. As hostility to pagan literature diminished, they preserved many Greek and Roman classics that would otherwise have been lost. Useful conventions such as separating words and using punctuation were gradually adopted, and around 800 "small letters" (minuscules) were introduced (until then, all writing had been in capitals, or majuscules). Many books were objects of great beauty, with elaborate painted initial letters, ornamentation, and pictures. Famous examples of these illuminated manuscripts include such products of the Irish school as *The Book of Kells* and *The Lindisfarne Gospels*.

Monks also created original works. In Britain these included Bede's *Ecclesiastical History*, Asser's *Life of King Alfred*, and works of theology and hagiography. In many monasteries scribes compiled chronicles, ranging from bare year-by-year contemporary records to histories from the Creation and more imaginative flights such as Geoffrey of Monmouth's *Historia Regum. The Anglo-Saxon Chronicle* (actually several annals) began the English tradition to which Higden's *Polychronicon* belonged.

In the later Middle Ages the spread of secular literacy led to the appearance of scribes and illuminators who were not monks. And in the mid-fifteenth century the invention of printing from movable type, on paper, began a revolution that ended the role of the monasteries in the making of books—even if Higden's chronicle was one of the earliest works to be printed in England by William Caxton.

Below: The peak of the illuminator's art: a page from Saint Luke's Gospel in *The Lindisfarne Gospels*.

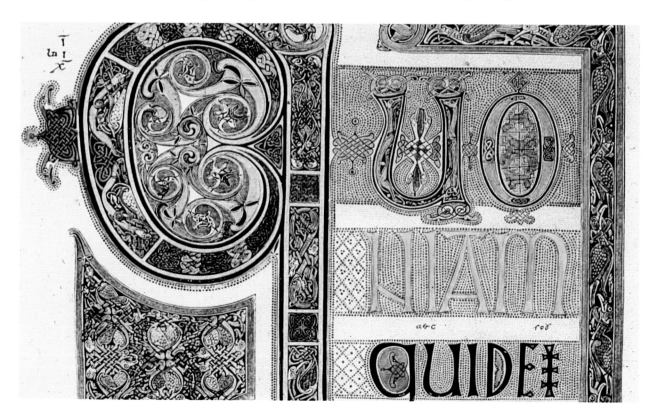

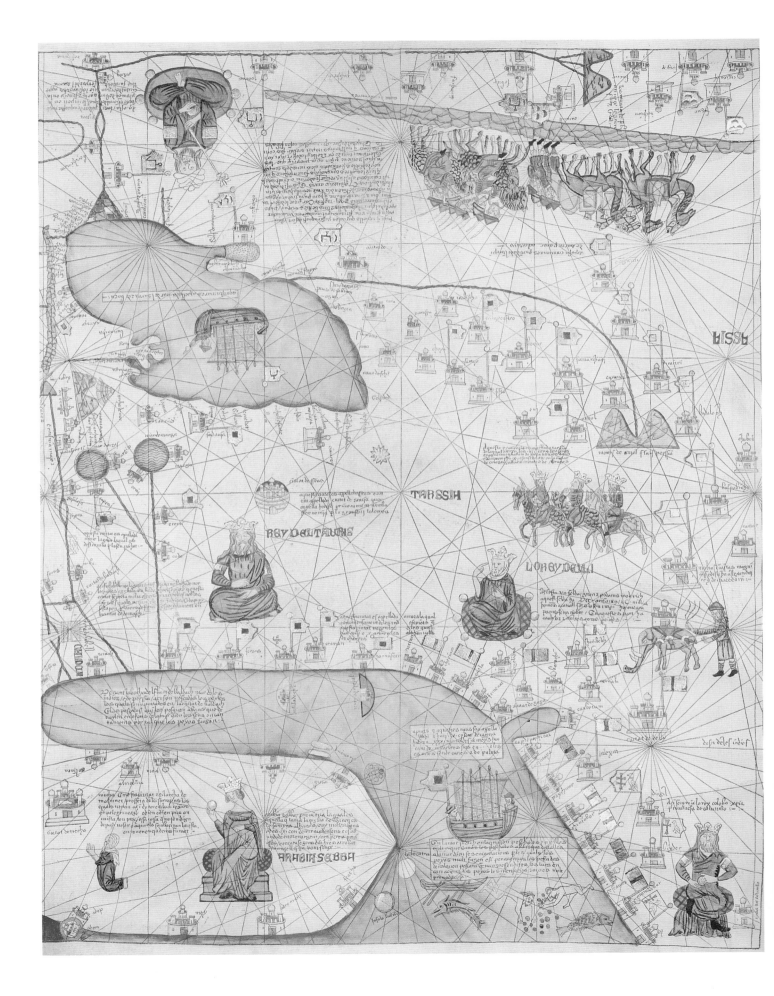

CATALAN ATLAS (ASIA)
ABRAHAM CRESQUES

This is a section from perhaps the most beautiful of all medieval maps, the Catalan Atlas of 1375. It was remarkably advanced for its time, combining pictorial images with generally accurate information, and traditional map features with novel items provided by contemporary travelers and traders.

Though always known as the Catalan Atlas, Abraham Cresques' masterpiece is actually a very large world map measuring 27 in. by 13 ft. (69 cm by 3.9 m), now consisting of twelve panels. Four of these are filled with cosmological and navigational lore, including a spectacular lunar calendar that makes it possible to date the atlas to 1375. The other eight panels form a long, rectangular map of the world by omitting the far-northern and southern regions. Though drawn in the style of a portolan chart, it was clearly not intended for use at sea but as a prestige item. It was presented by Pablo III of Aragon to King Charles V of France in 1381. Like a chart, the map must have been meant to be viewed on a table, from any side; hence it is "upside down" at the top. The panel opposite shows part of western Asia, taking in the region between the Persian Gulf and the Caspian Sea, northwest India, and part of the ancient caravan route across Central Asia. It is the first map to mention and make use of the writings of Marco Polo.

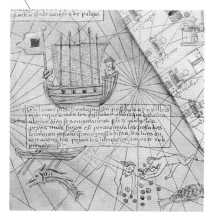

INDIAN SULTAN
The King, or Sultan, of Delhi. Cresques' information about India, and aspects of China not mentioned by Marco Polo, suggests that he had a wide range of contacts, especially among Arab traders.

THE POLOS' CARAVAN
This famous group of camels, drivers, and riders on horseback is often said to portray the Polos. The accompanying text is certainly adapted from Marco's writings, describing where to get supplies before crossing the desert and warning of its hazards.

THE QUEEN OF SHEBA
The kingdom of Sheba, represented by its legendary queen, is correctly placed in southern Arabia. The historical image is accompanied by a note saying the land had since become Muslim Arab territory.

PEARL DIVERS
Lively sketches of pearl divers on the Indian coast, based on descriptions by Marco Polo. They are said to use magic spells to frighten away sharks. Above sails a Chinese junk, much seen in the Indian Ocean during this period.

MARCO POLO

Much of the information about Asia on the Catalan Atlas was derived from Marco Polo's account of his adventures over eighty years before. At that time the East had been open to Western travelers—but only briefly. Consequently, the narratives of Polo and a few others remained the basis of European knowledge—and maps—until the sixteenth century.

In 1206 Genghis Khan united the Mongol tribes and embarked on a career of conquest. He and his successors created an empire that stretched across Asia from European Russia to China. The Mongols were feared for their savagery, but, paradoxically, the existence of a single Asian empire, replacing many warring states, meant that licensed travelers could travel its length in safety. The "Pax Mongolica" made it possible for Europeans to see the wonders of the East at first hand.

However, curiosity was not a prime motive for travel. The first visitors to the East were intrepid friars hoping (in vain) to convert the Great Khan and his people. They reached the Khan's Mongolian capital, Karakorum, but did not go on to China. That was reserved for a family of Venetian merchants, the Polos. On their way home after profitable trading in southern Russia, the brothers Niccolò and Maffeo Polo made a detour to Bukhara to avoid a war zone. There they met an envoy who promised them a warm welcome at Cambaluc (Khanbalik; modern Beijing), the Chinese capital of the Great Khan, Kublai.

The Polos meet Kublai Khan
The Polos took up the challenge and, as exotic strangers, were well treated by Kublai Khan, a shrewd and intellectually curious ruler. Interested in their religion, he dispatched them to the West with a request that the pope should send missionaries to argue the Christian case with followers of other creeds.

The Polos reached Italy in 1269, were supplied by the pope with two friars rather than the hundred that Kublai had actually asked for, and by 1271 were on their way back to China. The friars were quickly frightened off, but the Polos had acquired an

Above: This oriental-style statuette is believed to show Marco Polo during the seventeen-year stay in China that made him one of Europe's main sources of information about East Asia.

important fellow traveler in Niccolò's seventeen-year-old son, Marco.

Their journey took three and a half years, during which time Marco observed oil spouting at Baku, crossed Persia, went over the Pamirs, and took the southern branch of the ancient Silk Road into China. Once again the Polos were well received by Kublai, and by his own account Marco became a particular favorite, employed on special missions that took him all over the empire. This is perfectly plausible, since Kublai and the Mongols were conquerors, unwilling to trust their Chinese subjects with important missions.

Later Marco described the wonders of north and south China, Cathay and Manzi. (In time Europeans wrongly labeled the entire empire Cathay, or Chataio.) Surpassing anything in Europe were Kublai's triple-ringed palace at Cambaluc, parades featuring 5,000 elephants, and "the most splendid city in the world," Quinsai (Hangzhou). The former southern capital, Quinsai boasted canals, lakes, bridges, and pavilions, in addition to places of entertainment that catered for its million-strong population.

In 1291, after the Polos had served him for seventeen years, Kublai reluctantly allowed them to leave China. On their way home they carried out a last mission, escorting a Mongol princess to Persia, where she was to be married.

After his return to Venice, Marco would probably never have put his travels on record if he had not been captured during a war with Genoa. In prison he dictated his *Travels* (or *Description of the World*) to a fellow captive, one Rustichello, who happened to be a writer of romances. Rustichello certainly threw in materials of his own, complicating any evaluation of "Marco's" statements. Because of some inconsistencies and omissions in the text it has even been argued that Marco never reached China, retailing his memories of what he had been told by his father and uncle. Irrespective of their source, however, Marco's *Travels* did contain a great deal of accurate information. And this was all the more important because, after the death of Kublai in 1294, rival khanates were soon making war and the route across Asia was effectively closed. For well over two hundred years, writers and mapmakers would take their notions about Central and East Asia from the pages of Marco Polo.

Above: After their long journey Marco Polo and his companions kneel before the Mongol emperor, Kublai Khan, in this illustration from a late medieval collection of stories of marvels.

CATALAN ATLAS (AFRICA)
ABRAHAM CRESQUES

Cresques' treatment of Africa confirms the relative sobriety of his general approach
to mapmaking. Though some elements of medieval fantasy do appear, he invented
little and did not represent monsters as some of his contemporaries did, preferring to fill
the blank spaces on his map with evocative sketches.

Compared with the detailed treatment of the Mediterranean coastline, the rest of Cresques' map has few markings and many pictures and captions. He was not without sources of information, however. He was aware of the desert-dwelling Tuareg tribes, and of the caravan trade in salt, ivory, and, above all, gold that linked sub-Saharan Africa with the Muslim north. The wealth of the distant West African kingdoms is represented by the figure of Mansa Musa

(Cresques garbles his name into Messe Melly), a king of Mali who made a pilgrimage to Mecca in 1324, laden with gold. Cresques drew the Atlantic coastline and island groups with greater accuracy than his predecessors. By contrast, he underestimated the north–south distance to the gold-rich kingdoms, and his map does not reflect the immensity of the Sahara. He was shrewd enough to end his map where the half-known world ended and pure speculation began.

LOST EXPLORER
The little ship commemorates an
expedition led by the Catalan
mariner Jaime Ferrer. He set out
in 1346 to find the legendary "river
of gold" flowing into the Atlantic,
and was never seen again.

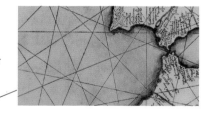

SAILORS' FRIEND
On both sides of the Pillars of
Hercules (Strait of Gibraltar), the
portolan chart style is kept up, with
carefully delineated coastlines, dense
lists of coastal ports, and rhumb lines.

CAMEL RIDER
A desert-dweller mounted on his
camel, with an exaggeratedly
ornate set of tents behind him.
His veil, and the accompanying
text, make it clear that Cresques
was representing a Tuareg warrior
or trader.

LEGENDARY KING
Messe Melly, described on the map as
"the noblest and richest king in all
the land." He sits enthroned,
displaying an orb or nugget of gold,
the source of his enormous wealth.

IMAGINING THE DARK CONTINENT

Even in Cresques' time, European ideas about Africa south of the Sahara were shaped mainly by medieval imaginings. While images of monstrous races gradually gave way to visions of gold, and later still to the harsh realities of the slave trade, to Europeans, Africa remained "the Dark Continent."

North Africa belonged to the Mediterranean world from ancient times, and eventually formed part of the Roman Empire. Even after its conquest by Islam in the seventh century, it remained not entirely unfamiliar ground. The Sahara barred the way south to all but the intrepid few, however, and consequently sub-Saharan Africa became a land of mystery. Medieval mapmakers delighted in populating the region with monstrous beings, such as the man-eating Anthropophagi, the four-eyed, bow-wielding Maritimi, and the cave-dwelling Troglodytes. On the Psalter and Hereford *mappaemundi* (pages 28 and 40), these and other creatures are lined up round the edge of the continent, which clearly served as a dump for

unused fantasies. When the circular world plan of the Middle Ages was abandoned, mapmakers followed Ptolemy in extending southern Africa to link up with the Far East or, like Cresques, avoided the issue.

Meanwhile, cross-Sahara contacts had become a little more frequent from the fourth century, as camels, famed as "the ships of the desert," became more numerous in North Africa. Arabs traded with desert caravans and converted much of West Africa to Islam. Large kingdoms or empires arose (Ghana, Mali, Songhai), with enough wealth to make Timbuktu a great center of Muslim learning. Though distant rumors of such distinction reached Europeans, they marveled even more at tales of gold in vast quantities, brought across the desert from some unknown source on the west coast. The lure of Africa now became great, but all sorts of fears persisted—not only of monstrous races but also of the tropical sun, which might make the sea (or the traveler's blood) boil.

Below: A caravan of camels crosses India's Rajasthan Desert in the cool of the evening. Fabled for its hardiness, the camel enabled the peoples of North Africa to trade across the Sahara centuries before Europeans managed to cross the great desert.

Above: Famed among Europeans as one of Africa's fabled cities of gold, Timbuktu, which is now in Mali, was founded by nomads in the eleventh century. It did indeed grow wealthy through trade, and became a great center of Islamic learning.

Looking for Prester John

For centuries Europeans knew as little about Asia as they did of Africa, despite some familiarity with biblical sites. In the late twelfth century, the travels of Marco Polo and others revealed that Asia, however full of wealth and wonders, was short on the magical and monstrous. Fantasies now tended to take refuge in Africa, among them the curious case of Prester John.

Tidings of a Christian king-priest (hence Presbyter, or Prester, John), the ruler of a vast "Indian" kingdom, reached the West in the mid–twelfth century. One of the motives for despatching missionaries to the Mongol-ruled East was to negotiate an alliance with Prester John against Islam, although it was soon clear that the Great Khan was not a Christian potentate.

Once John had failed to materialize in Asia, Europeans promptly transferred their hopes to Africa and identified his mighty realm with a real but less mighty land, the dimly known, long-isolated Christian kingdom of Ethiopia. In the fifteenth century, when the Portuguese prince Henry the Navigator began sending ships down the west coast of Africa, he hoped that they would find Prester John (and much gold). The legend kept such a hold on the European imagination that, as late as 1558, Diogo Homem's map of Africa (page 84) shows the seemingly ageless king, enthroned, in Ethiopia.

The exploits of Portuguese mariners in rounding the Cape and crossing to India began a new era. European fire- and sail-power, and the growth of the slave trade, marked the beginning of a self-justifying belief that Africans were inferior. The interior, largely unknown until the nineteenth century, became to European eyes an oppressive and sinister, rather than marvelous, Dark Continent.

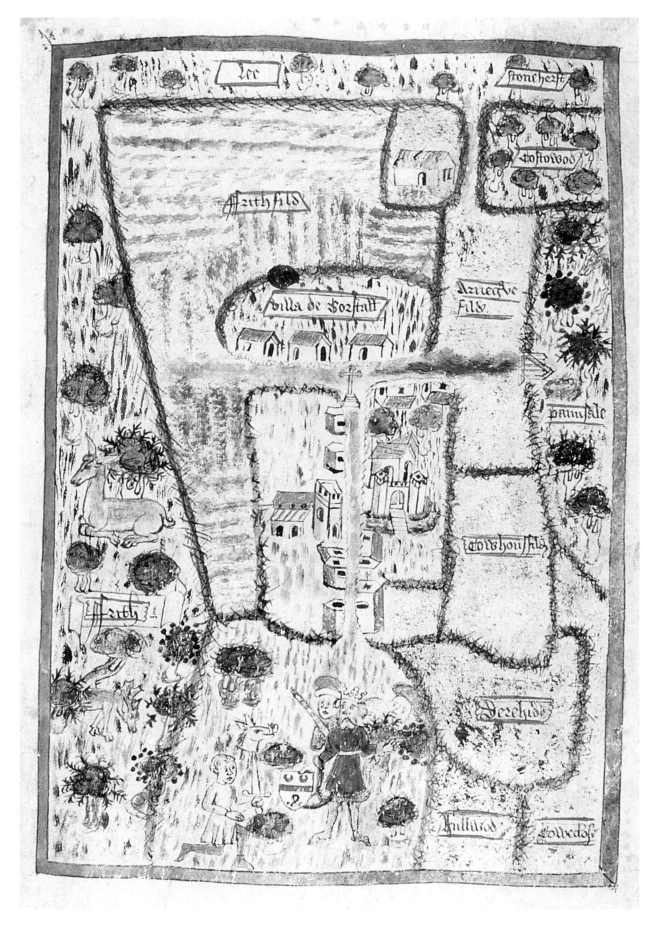

BOARSTALL MANOR

This early plan of the English village and manor of Boarstall gives a fascinating glimpse of a fifteenth-century community as its lord wished it to be put on record. Though much concerned with his prestige, the map pictures the settlement with surprising accuracy.

The map was drawn in 1444 for Edmund Rede, the lord of the manor, and is one of the earliest surviving village plans. It shows the moated manor house, the church, the village, Stonehurst Common, open fields for cultivation, and the extensive surrounding woodlands. Rede's title to the manor is asserted by the foreground scene in which Edward the Confessor grants it to Rede's ancestor, Niel. The entire map breathes the feudal spirit—a spirit that was in fact decaying as peasants ceased to be serfs and became tenants paying rents. Although the drawing is not detailed enough to be of much service in resolving property disputes, it is accurate as far as it goes. This has been demonstrated by investigations that have matched the ridge and furrow patterns on the map with those that can still be identified on the site. The surrounding woodland, and especially the prominent deer, assert Rede's importance. Stag-hunting was the king of sports, once literally confined to royalty but by this period open to all gentlefolk; whereas villagers were still liable to severe penalties even for snaring rabbits. But poachers still accepted the risks, and a poem of 1380 celebrates the thrill of killing a hart at dawn.

VILLAGE HOUSES
The village of Boarstall, showing the cross and the church and manor below it. It is close to the fields in which the inhabitants would have gone to work.

GRANT
Edmund Rede's ancestor kneels before Edward the Confessor to receive the title to Boarstall Manor, granted as a reward for his courage in killing a large boar whose head he is presenting to the king.

MANOR
Boarstall Manor is shown as a fortified dwelling protected by a twin-towered gateway and surrounding moat. The fourteeth-century gateway, with its hexagonal towers, is now a National Trust property.

CHURCH
The village church is a substantial building; the size of church and manor served to remind the villagers in their humble dwellings of the might of the sacred and secular authorities.

THE MEDIEVAL ACHIEVEMENT

To the modern eye, the images on the Boarstall Manor map suggest a changeless way of life based on small communities and a social order dominated by nobles. In reality medieval Europe was a much more varied and dynamic society, innovating and continuing to develop even in the face of calamitous setbacks.

After the western Roman Empire collapsed in the fifth century A.D., a new Christian European society struggled to establish itself while wave after wave of invaders and raiders battered at the continent. Towns decayed, and most people lived in small, self-sufficient rural communities. They were protected—and exploited—by the arms of local lords and their followers, who expected to be supported by the peasantry. In northern Europe this relationship evolved into the feudal system, with a hierarchy of lords and vassals and a peasantry reduced to serfdom. The Catholic Church, lavishly endowed, fitted into the system, and monasteries in particular became great landlords.

Europe survives the worst

The "system" was still young when, about A.D. 1000, it became clear that Europe had survived the worst. Signs of recovery and advance rapidly multiplied. Agricultural production increased, helped by a significant change in the climate that brought drier, warmer weather. The population rose sharply, towns revived, and more land was taken into cultivation. Internal colonization was intensified by an impulse toward monastic reform that led orders such as the Cistercians to reject easy living and found houses in remote places. Cleared, irrigated, and planted, these places increased European productivity.

Here and elsewhere, agricultural surpluses made it possible for specialized crafts to develop. As their products found their way to markets, they modified the self-sufficiency of village communities and gave a

Above: Despite agricultural innovations, farming remained backbreaking work. The whole village joined in during crucial periods such as the harvest, shown in this fourteenth-century illustration, with women and children helping the men in the fields.

Above: Blacksmiths at work in a village forge, as shown in an illuminated manuscript from the fourteenth century. The smithy was an essential part of any rural community, like the mill and the church.

renewed importance to money and merchants. A modest increase in trade and travel brought about some improvement in roads and a boom in bridge-building, although travel by water was still safer and, where possible, preferred. Population increase also led to a physical expansion of Western society, as people left overcrowded places to settle new land. Among them were the eastward-moving German colonists, who cut down the great forests as they advanced.

Pressure of demand acted as an economic stimulus. The manorial, or three-field system, based on the rotation of crops, proved effective in supporting a rising population. Among the most striking medieval innovations, especially important in northern Europe, were the introduction of water mills and the adoption of the heavy, iron-tipped plow, a better-designed horse collar harness, and the horseshoe. Not surprisingly, the central place of the blacksmith in village life and lore dates from this period.

In the twelfth century the growth of intellectual enquiry was reflected in the founding of universities. Philosophy became a subject of central concern, stimulated by translations into Latin of Aristotle and other ancient Greek thinkers. Their works had been preserved by Muslim civilization, with which Christian Europe came into violent but fruitful contact in Spain and Palestine. A wider knowledge of astronomy and mathematics, like the use of paper, also reached Europe via the Islamic world.

In some respects medieval civilization can be said to have reached its apogee in the thirteenth century, with the flourishing of the Gothic style in architecture and the achievements of figures as various as St. Francis of Assisi, Roger Bacon, and Marco Polo. The following century was one of calamities: population growth became unsustainable and caused famines; there were devastating wars; and above all the Black Death, a plague that spread throughout Europe from the East, killed about a third of the population.

The results were not entirely negative. The Black Death hastened the break-up of feudalism, creating a labor shortage that enabled peasants to improve their conditions, exchanging serfdom for tenant farming. Marginal lands were no longer cultivated, increasing the yield per head, and no population explosion undermined prosperity. By the fifteenth century semi-industrialized cities in Flanders and Italy were creating new artistic forms, a full-scale revival of classical learning, the Renaissance, was under way, and gunpowder and printing had been invented.

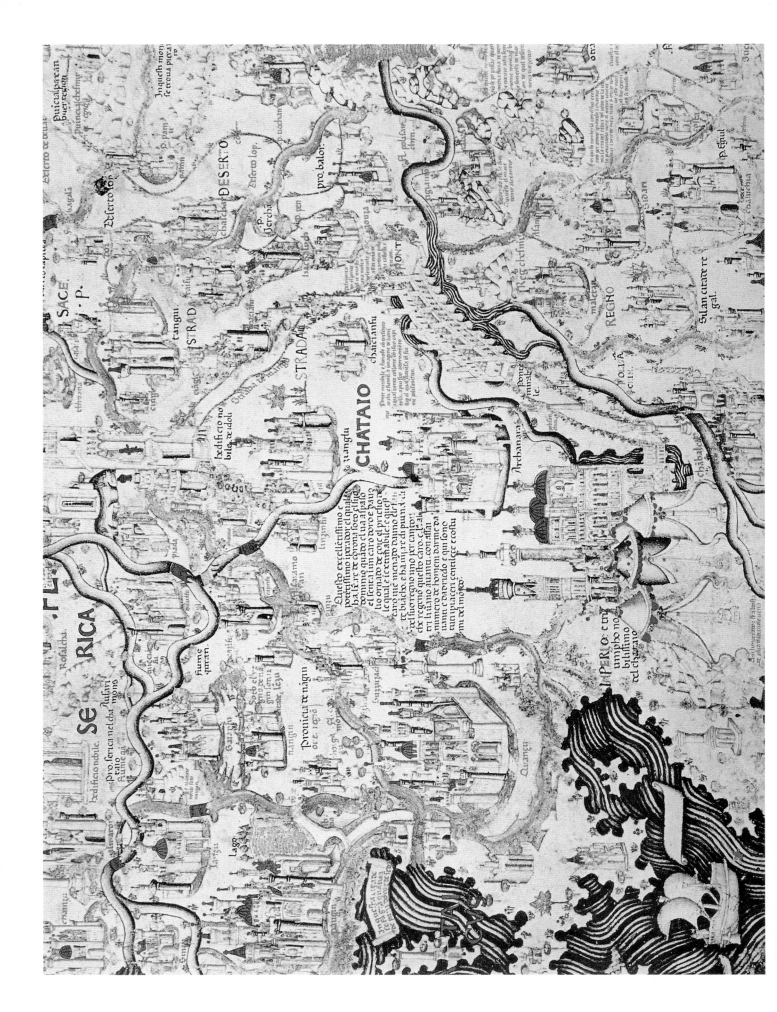

CATHAY
FRA MAURO

This image of Cathay, or China, is part of a world map made by Fra Mauro, a monk celebrated in his own time as a cartographer. While retaining the *mappamundi*'s traditional circular form and decorative charm, Mauro made a serious attempt to incorporate sound geographical information. Consequently, his work neatly bridged the medieval and modern worlds.

Fra Mauro was a renowned Venetian mapmaker, although neither of his earlier documented works, including a world map, has survived. His famous *mappamundi*, commissioned by Alfonso V of Portugal, was begun in 1457 and finished in 1459. The original has been lost, but fortunately a copy survives that was made soon afterward. Mauro had died by this time, and the copy was probably completed by his assistant, Andrea Bianco, a ship's

officer and chartmaker. Almost 6½ ft. (2 m) across, drawn on parchment and mounted on a wooden frame, it is crammed with notes and illuminated images. The south is at the top, so the Pacific coast of Cathay appears on the left. Information about the Far East was still based mainly on Marco Polo's *Travels*, but Mauro seems to have had another source, not available to his predecessor, Abraham Cresques, since China's river system is shown much more accurately.

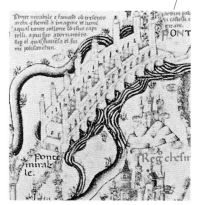

CAMBALUC
A Venetian palace and spacious tents symbolize Cambaluc, the capital of Chataio (Cathay) under Kublai Khan and his successors, although these Mongol emperors had long ceased to rule China by Mauro's day.

BEAUTIFUL BRIDGE
"The wonderful bridge" is clearly based on Marco Polo's description of a magnificent 24-arch structure about 9 mi. (15 km) west of Cambaluc: "There is not a bridge in the world to compare with it."

IDOLS
"A noble building of idols," according to the inscription, is evidently inspired by Marco Polo's frequent descriptions of places where "the people are great idolaters and bury their dead."

MAURO'S *MAPPAMUNDI*

Mauro's beautifully illustrated world map drew on influences as diverse as classical geography, the travel tales of Marco Polo, and then-recent Portuguese discoveries. With this work the *mappamundi* tradition effectively came to an end, having absorbed as much information as was compatible with its compact scheme of the three Old World continents.

In general appearance Mauro's work was still a *mappamundi*. Even its southern orientation was not notably unorthodox, especially in Italy, where contacts with the Islamic East were frequent and al-Idrisi had drawn maps four centuries earlier (page 24). However, Mauro displayed little of the schematizing mentality of the medieval cartographer. Accepting that Asia was larger than previously supposed, he placed Jerusalem farther west than its traditional position at the center of the world, though he did excuse himself by explaining the geographical realities while arguing that Europe's greater population density meant that Jerusalem was truly central—demographically.

The medieval tradition undermined

By the time Mauro created his *mappamundi*, medieval traditions were being undermined by the new Renaissance spirit, and ancient Greek and Roman texts—some only recently discovered or acquired by Europeans, sometimes through Islamic scholars and libraries—were becoming the new authorities. The *Geography* of Claudius Ptolemy, translated into Latin, had introduced new ideas that cartographers were beginning to copy. Mauro showed considerable independence of judgment, although he felt obliged to include a deferential note to explain why he did not use Ptolemy's parallels and meridians—lack of space. He refused to follow Ptolemy in making the Indian Ocean an inland sea, leaving open the possibility of a passage to the East around Africa. There was also no trace on his map of Terra Australis, the large southern continent whose existence had been proposed by Ptolemy and would be believed in by Europeans for centuries to come.

Like other fifteenth-century creators of *mappaemundi*, Mauro replaced the featureless contours of the genre with more accurate coastlines derived from portolan charts, even copying their tendency to overemphasize bays and headlands. However, there are no compass roses or rhumb lines on Mauro's map, and it was certainly not intended for use as a sea chart.

Mauro's Portuguese clients had up-to-date charts of their own that recorded their voyages down the west coast of Africa, and these were jealously guarded as state secrets. King Alfonso probably commissioned Mauro as a way of getting the best-available information about Asia and the Indian Ocean from a leading mapmaker working in Venice, the European city most closely in contact with the East. Though Mauro boasted that he had the use of many Portuguese charts, his map is accurate only as far as 12.20 degrees north, which the Portuguese had reached in 1446. The results of later voyages were kept from him, or he may even have been fed false information. On his map the African coastline curves away sharply to the southeast, perhaps because he had heard rumors of the existence of the Gulf of Guinea.

Mauro probably had reliable contacts in the Coptic Church of Abyssinia, members of which are known to have visited Venice. Though individual features of his depiction of Africa are quite accurate, he expanded Abyssinia so that it filled much of the continent. The presence of Arab names on the East African coast suggests that he also had some Islamic contacts.

The outline of Asia owes a good deal to Ptolemy, but much of the information about the East comes from Marco Polo and a more recent Venetian traveler, Niccolò de'Conti, who had spent twenty-five years wandering in India, Burma, and the East Indies (Indonesia). Mauro's representation of India has obvious shortcomings, but Sumatra is named for the first time on a map and Java is identified as the center of a great international spice trade. Most intriguing of all, an island just north of Java is named as Zimpagu, which, though badly misplaced, must be the first appearance on a European map of Cipangu (Japan).

Right: A detail of a Chinese landscape painting from the fifteenth century, with atmospheric mountains and streams.

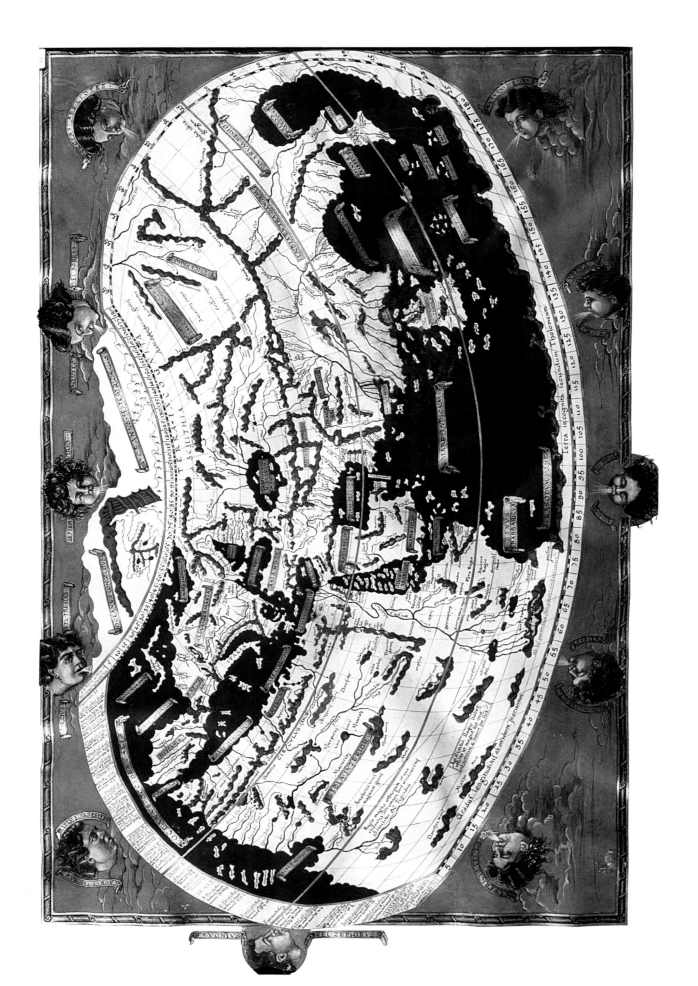

PTOLEMY'S WORLD MAP
JOHANNES SCHNITZER

In the fifteenth century, the works of the ancient Greek geographer Claudius Ptolemy, forgotten in the West for a thousand years, were translated into Latin and enthusiastically studied. Among the many new editions of Ptolemy's *Geography*, a German version published in 1482 is notable for its twenty-seven strikingly bright and bold maps.

By the 1470s Ptolemy's work had a relatively wide influence through the new medium of printing with movable type. The most distinctive feature of the 1482 German edition of Ptolemy, which included the world map illustrated here, was that its maps, unlike those in other editions, were printed from woodcut blocks rather than engraved copper plates. More subtle effects were possible with copper engraving, but the strong colors and boldness of woodcuts could also be artistically satisfying in the hands of a master. The master here was woodcut artist Johannes Schnitzer, who worked from maps drawn by Dominus Nicolaus Germanus, a Benedictine monk. Germanus partly updated his Ptolemaic map, but essentially it shows the world as known to antiquity. It is trapezoidal in shape and has been drawn on a projection with lines of latitude and longitude, based on Ptolemy's coordinates: such sophisticated concepts were to have considerable significance for the future of cartography.

THE FORTUNATE ISLES
The representation of the west coast of Africa shows no awareness of the fifteenth-century Portuguese voyages of discovery. The only "modern" feature is the Canary Islands (Fortunate Insule, or Fortunate Isles), discovered around the 1330s.

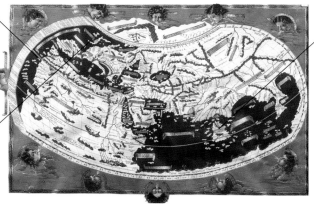

GIANT ISLAND
A huge Sri Lanka (Taprobana) appears below a truncated India. Africa extends all along the southern edge of the map, joining the Far East, and the Indian Ocean is shown as an inland sea.

STRETCHED SCOTLAND
The strange northeastward elongation of Scotland is one of Ptolemy's more curious errors, copied by Schnitzer. Scandinavia, unknown to Ptolemy, has been added, along with other far-northern features that appear beyond the edge of the map.

THE INFLUENCE OF PTOLEMY

From the fourteenth century onward, Italy, and later the rest of Europe, underwent a great intellectual revival known as the Renaissance. Much of its energy was derived from the rediscovery of ancient writings and works of art, especially those of the Greeks. Among the most influential were geographical and astronomical texts by Claudius Ptolemy of Alexandria.

So little is known of Ptolemy's life that even his dates of birth and death are disputed. His major works were written somewhere around the middle of the second century A.D. He spent most of his life at Alexandria in Egypt, which was at that time a great center of Greek culture with the most famous library in antiquity. Though Ptolemy was most celebrated as a geographer and astronomer, his contributions to learning extended to optics, mathematics, and other scientific subjects.

During the Dark Ages that followed the fall of the Western Roman Empire, contact between Europe and the Greek East was at best sporadic, and works such as Ptolemy's were unknown. This was not true of Byzantium, however, or of the Islamic civilization that took over much of the Greek world. In fact, the translation into Arabic of Ptolemy's work on astronomy, the *Mathematical Compilation*, was so admired and used by Arab astronomers that it is still known by its Arabic title, the *Almagest* ("Greatest"). It reached Europe in this form in the twelfth century, when it was translated into Latin, the scholarly language of the West, by Gerard of Cremona.

Ptolemy's astronomical achievements included a catalog of more than a thousand stars and their positions. Many of his explanations were of lasting value, but his belief that the Sun and the planets must move around the Earth, and in perfect circles, had unfortunate consequences. Since his theories did not fit in with what he himself could see in the night sky, he devised an ingenious, if cumbersome, system involving lesser orbits (epicycles) to reconcile theory and observation. The Ptolemaic system acquired immense authority and, backed by the authority of the Church, became a bar to astronomical progress.

Other previously unknown Greek works reached Europe as Renaissance scholars acquired manuscripts from a rapidly declining Byzantium. The *Geography* became accessible about 1406, when Jacobus Angelus translated it into Latin. Many manuscript copies were made within a few years, and with the spread of the new invention of printing, published editions multiplied. The first, at Verona (1475), contained the text but no maps; then editions with maps followed at Bologna (1477), Rome (1478), Florence (about 1482), and across the Alps at Ulm (1482, reissued in 1486). There were many more editions, culminating in Mercator's famous 1578 edition, reprinted many times.

The *Geography* listed 8,000 places with their coordinates (that is, their latitude and longitude). It also discussed various possible projections, or methods of representing the curved surface of the Earth on the flat surface of a map. Ptolemy had no doubt that the

CL·PTOLEMAEO·ALEX·

PHAROS

Earth was a sphere, though he understandably believed that its inhabited areas were confined to Europe, Asia, and a truncated Africa. The maps that accompanied the *Geography*, though based on Ptolemy's coordinates, may not have been part of the original. Surviving manuscripts of Ptolemy's works are copies of copies (the earliest dates from about 1200), so the accompanying maps may have been added at any time over a period of a thousand years.

An inspirational work

The *Geography* had a tremendous impact on Europe. Its systematic treatment of the subject inspired a new seriousness, while the introduction of coordinates ultimately proved to be the key to accurate mapmaking. As with the *Almagest*, Ptolemy's authority became so great that some of his mistakes were copied on to new maps. However, the maps accompanying

Above: The lighthouse at Pharos, one of the seven wonders of the world; it stood at the entrance to the harbor at Alexandria from the third century B.C. until it was destroyed by an earthquake in the fourteenth century.

Far left: Ptolemy examines a planisphere in this sketch from the early modern period, when his works became highly influential.

the text were soon recognized as documents of historical rather than contemporary significance, and editions of the *Geography* began to incorporate modern as well as Ptolemaic maps—the Ulm edition, for example, has five of these additions. One error— Ptolemy's belief that Eurasia covered 180 degrees of the globe—was extraordinarily fruitful, convincing a mariner named Christopher Columbus that it would be a relatively simple matter to sail across the Atlantic to a not-too-distant Asia.

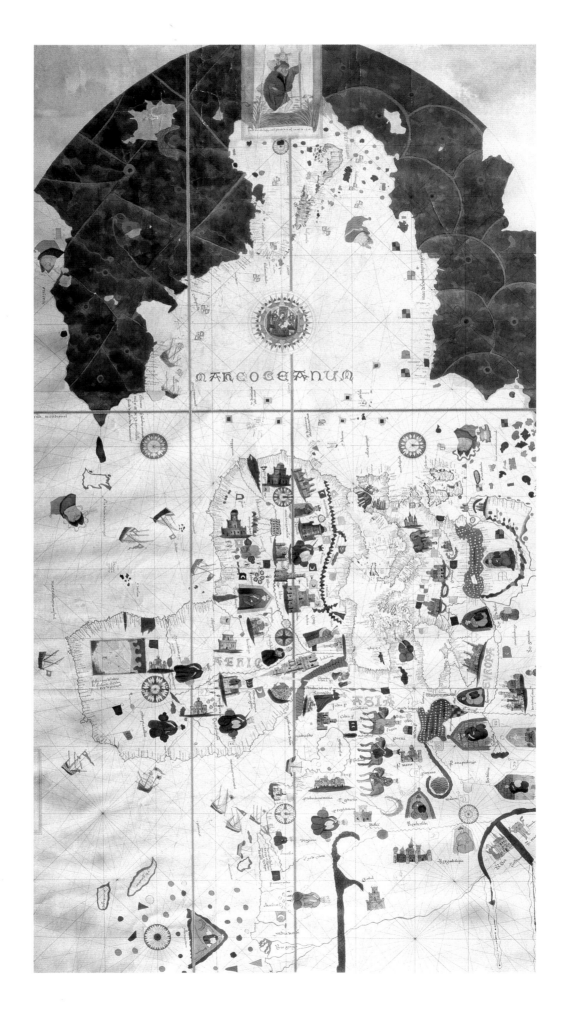

WORLD MAP
JUAN DE LA COSA

The first world map to feature Columbus's epoch-making discoveries was made by one of the Spanish mariners who accompanied him in 1492. Juan de la Cosa's map of about 1500 was remarkably up to date, showing the expanding world as it had become known to Spanish, English, and Portuguese explorers by the end of the fifteenth century.

When Christopher Columbus left Spain on his first transatlantic voyage in 1492, the master of his largest vessel, the *Santa Maria*, was one Juan de la Cosa. The next year, de la Cosa sailed with the second expedition. Remaining active in the Caribbean after the end of his association with Columbus, he took part in a further five expeditions, mainly devoted to exploring the northern coast of South America. He died in 1509 from the poisoned arrows of hostile Native Americans in Venezuela. De la Cosa's map reflects his detailed knowledge of the Caribbean, showing all the islands discovered by Columbus, as well as Jamaica, Puerto Rico, and Cuba (which, unlike his former master, de la Cosa correctly depicted as an island). Evidently in touch with English and Portuguese sources, he gives a good account of the coasts of North America and western Africa, but imagination—or misinformation—plays a large part in his pictures of Brazil, the eastern coast of Africa, and the relationship between the Americas and Asia.

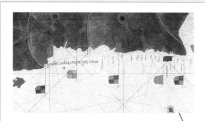

ENGLISH STANDARDS
The presence of five English standards, and the accuracy with which the American coastline is drawn, prove that de la Cosa was acquainted with the discoveries made by John Cabot in 1497.

MYTHICAL BEINGS
Medieval habits persist in the depiction of fabulous creatures: an animal-headed being and a headless "human" with its face in the middle of its torso suggest the biblical Gog and Magog.

TRAVELERS' SAINT
St. Christopher, patron saint of travelers. The placing of the image enabled de la Cosa to avoid committing himself to charting the existence of a passage to Asia in this area.

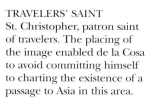
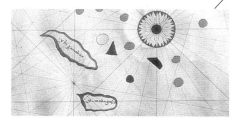

ISLANDS AWAY
Compass roses and rhumb lines bear witness to the durability of the portolan chart tradition. The islands of Zanabar (Zanzibar) and Madagascoa (Madagascar) have drifted far from the east coast of Africa.

A SPANISH VIEW OF THE WORLD

A secret from the past, Juan de la Cosa's world map was unknown for more than 300 years until its chance recovery. This lovingly drawn and colored manuscript is a unique document that brings us close to Christopher Columbus and pictures the world near the beginning of the great Age of Discovery as it appeared to a particularly well-informed individual.

D e la Cosa's map was found only in 1832, in a Parisian antique shop, by Baron Walckenaer, then serving as the Dutch ambassador to France. After his death in 1859 it was acquired by the Queen of Spain, and it is now displayed in Madrid's naval museum. Apart from the rough sketch of the coast of Hispaniola by Columbus—a sketch that modern satellite photography proves to be remarkably accurate—there are no other contemporary illustrations of comparable historical interest to this world map. Despite some deterioration, it is still a finely drawn and elegantly colored object.

In its present form, the parchment measures about 38 by 72 in. (96 by 183 cm). According to the latest academic thinking, it is a copy of de la Cosa's original, but a very early one. It appears to have been made somewhere between 1502 and 1510 from an original created as early as 1500. This is in fact the date given under the image of Saint Christopher, in a caption that also asserts de la Cosa's authorship. Inconsistencies in the map have led to controversies, with some scholars opting for a later date than 1500. However, the absence of information that would have

Above: The marriage of Columbus's sponsors, Ferdinand of Aragon and Isabella of Castile, in 1469 united the Christian kingdoms of Spain and gave the final impetus to the *reconquista*, the campaign to destroy Moorish power in the Iberian peninsula.

supporting his belief that Cuba was in fact a peninsula, linked to the Asian mainland. De la Cosa was among those who signed this eccentric attempt to make reality correspond to Columbus's dreams, possibly as a way to convince the monarchs at home of the success of the expedition. When de la Cosa came to produce his own map only six years later, it correctly showed Cuba as an island, suggesting that by 1500 (when Columbus was still alive) the truth was already known.

Cautious cartography

De la Cosa made the Americas disproportionately large in relation to the Old World, but in some important respects he was more cautious than most contemporary cartographers: A conveniently placed image of Saint Christopher obscures Central America, and with it the question of whether, as some believed, a sea passage existed in the area. Eastern Asia is treated with a vagueness that leaves open the relationship between Columbus's discoveries and the Far East, which Columbus himself fervently believed to be part of the same continent.

For all its modernity, de la Cosa's map combined the recording of the new discoveries with elements from two older cartographic traditions. The outlook of medieval world maps lived on in little pictures that mingled biblical, historical, and fabulous elements: the Three Kings cross Asia bearing gifts, the Tower of Babel still rises beside Babylon, and Prester John continues his centuries' long reign in Africa. Simultaneously, the map also follows the portolan chart tradition in its accurate delineation of the Mediterranean, the naming of coastal locations (neatly tucked into the land side even where the drawing of the coastline is largely speculative), and in the presence of compass roses and rhumb lines.

De la Cosa's map, primarily reflecting Spanish experience, was followed by the "Cantino planisphere" of 1502 from Lisbon, recording, among other Portuguese achievements, Pedro Álvares Cabral's discovery of Brazil in 1500. From this time onward, in manuscript and (from 1506) in print, the Americas and the expanding world of the East continued to challenge and puzzle cartographers.

been available to de la Cosa just a few years later than 1500 (including the results of his own voyages) strongly supports the current dating.

Putting Cuba on the map

De la Cosa's treatment of Cuba is of particular interest because of its close connection with a story about Columbus. During his 1494 voyage, the admiral sailed for some distance along the south coast of the island. Increasingly difficult conditions forced the expedition to turn back, but before doing so, Columbus insisted that his men put their names to a document

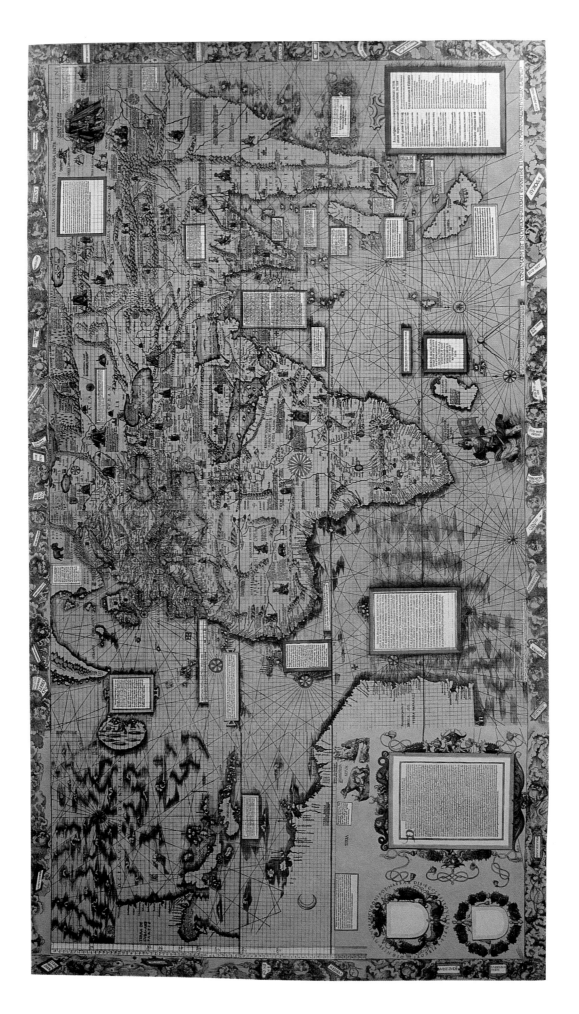

CARTA MARINA
MARTIN WALDSEEMÜLLER

The Carta Marina is one of several outstanding maps made by Martin Waldseemüller, who worked in apparent isolation in the Vosges mountains but managed to be well informed about the world-expanding discoveries of his time. The Carta Marina combines the functions of world map and sea chart in an unusual fashion.

Waldseemüller was probably born near Freiburg in Germany. He became a priest but little is known of his movements until 1506, when, aged about 36, he arrived at St-Dié in northeastern France. Evidently experienced in design and painting, he was welcomed into a scholarly circle patronized by the Duke of Lorraine. Over the years, primed by Italian and Portuguese contacts, Waldseemüller produced a world map (1507) that was crucial in naming America, a new edition of Ptolemy's *Geography* (1513), and the Carta Marina. This "Sea Chart of Portuguese Navigation" is a large (twelve-sheet) wall map; the only surviving copy is bound with Waldseemüller's 1507 world map. It is unusual in combining a sea chart's rhumb lines and coastal place names with a wealth of detail and imagery in the old *mappamundi* tradition. If Iberian exploration supplied the outline of the Americas and new African place names, the information given about Asia continued to rely on the writings of Marco Polo.

CROWDED EUROPE
Exploding rhumb lines off southwest England seem to hint at the unlimited possibilities revealed by the great voyages of exploration. Significantly, no exotic images are needed to fill the crowded map of Europe.

MIGHTY MONGOL
A Great Khan seated in his tent. This is the only figure approaching the size of the Portuguese monarch at the bottom of the map. However, even the mighty Mongol cannot match King Manuel's stature.

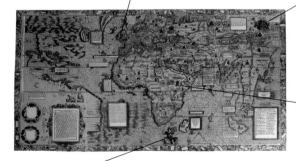

CAUTION PREVAILS
Though sometimes credulous, Waldseemüller was also cautious. The legendary priest-king Prester John appears, but it is left an open question whether his kingdom is here, in Ethiopia, or, as older traditions insisted, in India.

MASTER OF THE SEAS
A tribute to Portuguese achievements. A large figure of King Manuel I is shown astride a dolphin, a traditional image implying mastery of the watery element.

NAMING AMERICA

Though he lived a quietly studious life in old Europe, Martin Waldseemüller unwittingly usurped the privileges of the New World explorers by naming their discovery. The continent should have been called Columbus or Columbia; instead, through a combination of accident and misinformation, the Florentine adventurer Amerigo Vespucci had immortality conferred upon him.

Above: Amerigo Vespucci uses an astrolabe to take readings from the Southern Cross. Most contemporaries gave Vespucci rather than Columbus credit for the discovery of the New World.

Vespucci was much employed in Spain and Portugal and was a friend of his fellow Italian Christopher Columbus. He certainly took part in voyages to the New World, though his exact role is obscure. Scholars are now inclined to believe that he made two voyages, in 1499–1500 and 1501–1502, rather than the four that contemporaries accepted, including one in 1497–1498 that would have made Vespucci, not Columbus, the first European to reach the South American mainland.

Columbus's account of the islands he found on the other side of the Atlantic was not particularly inspiring. To contemporaries, Vespucci seemed a much more exciting character on the basis of his supposed four voyages, his firm conviction that South America was a "new world," and his talent for self-publicizing letter-writing. It was easy to imagine that Vespucci, and not Columbus or other early explorers, had been the driving force in the great discoveries.

Stealing Columbus's thunder

When the Duke of Lorraine received a copy of one of Vespucci's letters and some charts of the New World, he and his circle responded with enthusiasm. Work on their current project, a new edition of Ptolemy, was suspended in favor of designing a world map that would take in the wonderful contemporary discoveries. In 1506, as part of the preliminaries, Waldseemüller produced a set of twelve gores, the segments that make up a globe, labeling the new southern continent "America."

The following year he gave reasons for his choice in the *Introduction to Cosmography*, a text that explained the fundamentals of geography and astronomy. As well as admiringly quoting Vespucci's four voyages, Waldseemüller wrote that, since Americus Vesputius (the Latin form of Vespucci's name) had discovered the fourth continent, it should be called America, a feminine form of his name, coined by analogy with Europa and Asia.

Even more important was the fact that "America" was used on the great world map by Waldseemüller that accompanied the *Introduction*, complete with prominent portraits of Ptolemy and Vespucci as the supreme authorities on the Old World and the New. Very large—its twelve sheets measure 52 x 93 in.

(132 x 236 cm)—Waldseemüller's 1507 world map was a masterly, heart-shaped woodcut. It was also the earliest printed map to show the Americas as a landmass definitely separated from Asia, although Waldseemüller was apparently unable to make up his mind about the rumored strait passing through Central America, putting it in the main map but omitting it in the inset, a then-novel double-hemisphere representation of the world.

Waldseemüller's text and map were widely admired, purchased, and imitated; and "America" took hold. Meanwhile, Waldseemüller evidently acquired fresh sources of information that led him to revise his opinions. In 1513 he published the long-projected new edition of Ptolemy with twenty "modern" maps to supplement those derived from Ptolemy's writings. On Terre Nove (New Lands) "America" has gone from South America and two lines of script acknowledge that "This land, with its adjacent islands, was discovered by Columbus." In the introduction to the map, Waldseemüller credited "the Admiral" as his source. Since only Columbus was referred to in this way, it is supposed that the discoverer of America himself was Waldseemüller's contact, and Terre Nove is usually called "The Admiral's Map."

"America" was also absent from Waldseemüller's Carta Marina of 1516, but the cartographer's change of mind came too late. Numerous new mapmakers and imitators had committed themselves to the term, and the New World was fated to remain "America."

Above: The printing press—this example dates from 1528—and the rapid dissemination of books enabled Vespucci to maximize the effectiveness of his talent for publicizing his exploits.

Left: The title page from the account of Vespucci's four voyages, published in 1507. Although his claims to have discovered the New World were exaggerated, Vespucci was an accomplished mariner who had crossed the Atlantic and explored South America.

TENOCHTITLÁN
(MEXICO CITY)

Tenochtitlán was the capital of the Aztec empire, doomed to be destroyed by a
Spanish expedition to Mexico led by Hernán Cortés. This Spanish map is the best
surviving record of a city whose size and beauty caused the conquistadors who first saw it
to ask themselves whether they were dreaming.

Cortés captured Tenochtitlán in August 1521
and at once began razing the city. When he
sent this map, with an account of his exploits,
to the Spanish king in 1524, most of what it showed
had already disappeared. The plan, a map of the Gulf
of Mexico, and Cortés' dispatches were published in a
Latin translation later the same year in Germany.
Cortés' map captures something of the combination
of blues, greens, and whites that confronted the
Spanish when they looked down on Tenochtitlán.

Like Venice, it was an island city linked to the
mainland by artificial causeways. On its fringes floated
artificial gardens encased in wicker. Further in,
stepped pyramid temples rose above tight-packed
white-plaster or pumice-red houses. The canal-streets
and the surrounding lake were busy with canoes. In
the center of the island stood a great plaza holding
the royal palace and the most important temple,
where huge numbers of human sacrifices were offered
up to propitiate the Aztec gods.

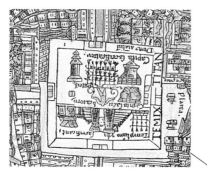

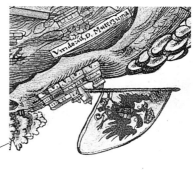

WATER ENGINEERING
The Aztecs' technology, though
simple, was harnessed skillfully to
their needs. The fence represents
the dyke built by Montezuma I
to dam the lake, which was
capable of flooding during
the rainy season.

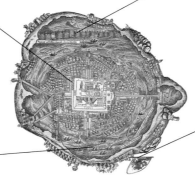

THE CITY'S SACRED CENTER
The plaza or sacred enclosure in the
middle of the city, with Montezuma's
palace, the great temple of Teocali,
and a headless sacrificial victim. The
little enclosure at the top right is
Montezuma's zoo.

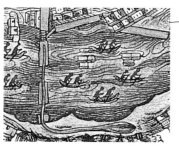

WATER SUPPLY
An aqueduct bringing fresh water to
Tenochtitlán from the springs of
Chapultepec; the first Aztec aqueduct
was constructed by Montezuma I half
a century before the Spanish arrived.

THE DOUBLE-HEADED EAGLE
These imposing buildings fly an
outsize flag with the double-headed
black eagle of Spain's ruling
Habsburg dynasty. Presumably this
was the palace from which Cortes
ruled Spain's new subjects.

THE CONQUEST OF MEXICO

Tenochtitlán was the capital of an Aztec empire that appeared to be at the height of its power and glory when it was destroyed by a few hundred Spaniards. The conquest of "New Spain" was an epic undertaking, but thanks to a combination of good luck and inspired leadership it was remarkably swift and complete.

H ernán Cortés was a Spanish adventurer who took part in the conquest of Cuba, for which he was rewarded with land, slaves, and the position of mayor of the capital, Santiago. At 33, still restless, he responded to rumors about the existence of a wealthy inland kingdom, mounting an expedition from Cuba in defiance of the governor. He gambled that a victory, particularly if it yielded substantial amounts of loot, would buy the Spanish king's retrospective favor.

Cortés set out with about 450 men. He also had a few cannons, horses, and dogs, all three of which were quite unknown to the mainland Indians. They helped make the Spaniards seem magical beings to friend and foe alike. Cortés sailed west along the Yucatan coast, where he acquired a native slave mistress, known as Dona Mariña, whose skill as an interpreter became vitally important to his success. Landing in Mexico in April 1519, he founded the town of Vera Cruz, quelled a mutiny, destroyed his ships so that there could be no going back, and began the march inland.

He was soon able to recruit allies. The Aztecs were only one among many Mexican peoples whom they dominated, and their wars and demands for tribute from their neighbors created great resentment.

Above: This symbolic Aztec map of their capital city shows the canal-streets of Tenochtitlán; at its center is the legendary eagle that is said to have guided the early Aztecs to the site of the city.

Left: Aztec myths told many stories of harsh, warlike gods, like this fierce guardian of Tenochtitlán. The city was a center for the ritual sacrifices that kept the gods contented.

Meanwhile the Aztec ruler, Montezuma II, sent rich gifts to the stranger while attempting to dissuade him from marching further inland. The emperor seems to have been in two minds about how to react to Cortés, willing to buy him off or trap him, yet also half-convinced that he must be the white god-king Quetzalcoatl, whom myths suggested would return to Mexico one day from the east.

Faced with a trap at the town of Cholula, Cortés launched a bloody preemptive strike that must have added to Montezuma's alarm. On November 8, 1519, Cortés entered Tenochtitlán, where Montezuma treated him as an honored guest. Realizing that he had walked into a gilded prison, Cortés kidnapped the king. Strangely passive, Montezuma cooperated with his captors, punishing Aztecs who attacked Spaniards and ordering Christian symbols to be placed on the altars where hundreds of thousands of victims had had their hearts ripped out, a ritual the Aztecs believed essential to ensuring the sun would rise each morning.

The Spanish take control

This uneasy situation lasted for months. In June 1520 the Aztecs finally rose against the Spanish and their compliant king. Montezuma was killed by the Spanish or his own subjects—each side accused the other—and Cortés and his men suffered heavy losses as they fled from the city along one of the causeways.

The Spanish retreated to the neighboring state of Tlaxcala, where they recruited thousands of men and built a small fleet of brigantines. The Aztecs' canoes were no match for these, and the Spanish, able to command the lake, besieged Tenochtitlán.

Ships, cannons, and cavalry triumphed. The last Aztec king, Cuauhtemoc, was captured and hanged and in August 1520 the city surrendered. Its temples and houses were razed, the rubble used to fill in the lake, and a new, Christian capital, Mexico City, built.

Cortés had been right to believe he would win the king's approval. He was appointed governor of New Spain, extending its frontiers far to the south. His experience of royal favor was short-lived, but the Spanish possessed the largest colonial empire in the New World for almost three further centuries.

Above: Hernán Cortés is greeted by Montezuma II on his arrival in Tenochtitlán in November 1519. The emperor was uncertain about how to treat the Spaniard, understanding the threat he posed to the empire but fearful of the consequences of resistance.

THE VALLARD CHART
THE FRENCH IN CANADA

This beautiful map combines the functions of a sea chart, a map of newly discovered lands, and a pictorial celebration of early French settlers in Canada. The most recently acquired information on it, including the course of the St. Lawrence River, derived from the exploits of the great French explorer Jacques Cartier.

The Vallard Chart was drawn on vellum (calfskin) in or just before 1547. It was made by an unknown cartographer at Dieppe, a well-known mapmaking center in Normandy, France. The Vallard who has given his name to the map appears to have been an early owner. The care taken over the chart's decoration makes it unlikely that it was intended for use at sea, although it is laid out in the traditional style of a portolan chart, with wind roses, rhumb lines, and place names mostly written at right angles to the coastline on the land side. West appears at the top, so that the map runs, left to right, from Greenland along the eastern seaboard to Florida. The northern regions are disproportionately large, reflecting Jacques Cartier's penetration of the Rio de Canada (St. Lawrence River) and his attempt to found a colony at Cap Rouge in 1541. The visual appeal of the Vallard map is greatly enhanced by its vivid pictures of the colonists and the Native American peoples they encountered.

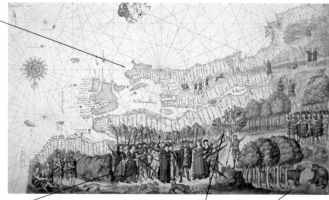

CARTOGRAPHIC CONFUSION
Cape Breton was already well known to the French sailors and fishermen, but the relationship between the island and Nova Scotia is one of the most confused features of the map.

COLONISTS
Settlers are depicted clustered around a leader. Their everyday European dress and the presence of women create an impression that is contradicted by the spears and halberds in the background and the men bearing muskets.

STRONG DEFENSES
A stockade, well furnished with cannon, protects the colonists who are shown in such confident poses close by. In reality, Native American attacks had made it impossible to maintain the settlement several years before the map was drawn.

LOCAL FAUNA
Fur-clad Native Americans are shown in pursuit of the local fauna, about which the artist had only very general ideas—as in the doglike black bears in the map's bottom right-hand corner.

CARTIER IN CANADA

After Columbus's discovery of the New World, its colder northern regions attracted European fishermen, but there was little interest in the mainland until the 1530s. Even then, the journeys of Jacques Cartier were driven by the dream of finding a route to the East. Unwittingly, Cartier blazed a trail that later French explorers and settlers would follow.

Newfoundland can claim to be the first place in the Americas settled by Europeans, since Vikings established a colony there, at L'Anse-aux-Meadows, in about A.D. 1000. It was abandoned after a few years, and Christopher Columbus made his Atlantic voyages of discovery—or rediscovery—by following a more southerly route.

Sauvage Iroquois

In 1497 an English expedition commanded by John Cabot made landfall in Canada (probably in Nova Scotia) before sailing on in search of a wealthier region of "Asia." English, Breton, and Norman sailors fished the waters at an early date, joined by the Portuguese after the voyages of the Corte Real brothers and João Fagundes. Fagundes may even have attempted to settle colonists from the Azores on Cape Breton Island in the 1520s, although the details of the episode are obscure, and the settlement, if it actually existed, vanished without trace.

Official French involvement began in 1524, when King Francis I sent Giovanni da Verrazano on a voyage that encompassed the east coast of North America from Cape Fear, North Carolina, to Newfoundland in search of the elusive sea passage to the East. In 1534 Francis tried again. This time the expedition was to be led by Jacques Cartier, a 43-year-old Breton navigator who had already sailed in Canadian waters and visited Brazil. His instructions from the king were "to discover islands and territories where, it is said, quantities of gold and other precious things are to be found."

An inland sea

Cartier's two vessels reached northern Newfoundland in May 1534 and then sailed down its west coast into the Gulf of St. Lawrence. They explored the south coast of the gulf and its islands, including Anticosti Island in the mouth of the St. Lawrence River, but they failed to locate the river itself. Nevertheless, on his return to France, Cartier was able to persuade the king that he had discovered an inland sea that, as the Native Americans had assured him, led to a land of great riches. Like other European explorers, Cartier treated the natives badly, kidnapping the sons of an Iroquois chief to serve as his interpreters. Yet at the same time, he believed the stories the natives told him.

On Cartier's second voyage (1535–1536), his hostages guided him into the St. Lawrence River as far as the village of Stadacona, close to the site of present-day Québec City, which became the explorers' base.

Left: The earliest native peoples encountered by the French included Iroquois, like this warrior, and Huron. Both initially suffered violence at the hands of the European newcomers.

Although the Native Americans tried to dissuade him, Cartier pushed on upriver as far as a village named Ochelaga, or Hochelaga. When he climbed a mountain that he named Mt. Royal (the origin of the name Montréal), he saw the Lachine Rapids—a formidable barrier, but one that was surely the gateway to Cathay (the East). It was already September, however, and Cartier returned to Stadacona for the winter. The unexpectedly harsh winter, the onset of scurvy, and deteriorating relations with the Native Americans led Cartier to return to France in May 1536.

By this time Francis had other preoccupations, and Cartier was not called on again until 1541, when a scheme to found a colony was mooted. The veteran explorer had to serve under a nobleman, Jean François de Roberval, but Cartier went on ahead and established a settlement at Cap Rouge, not far from his former base at Stadacona. During his wanderings Cartier discovered what he believed to be gold and diamonds, but after another painful winter, he and the colonists abandoned the settlement. Encountering de Roberval on Newfoundland, Cartier ignored an order to turn back and shipped away at night. Though his treasures turned out to be iron pyrites and mica, he was not punished, but he was never employed again by the French crown.

De Roberval soon gave up his colonizing plans, and the king lost interest in a region that had brought him no profits. Though Cartier had laid claim to Canada for France and shown the way down the St. Lawrence, half a century was to elapse before the French renewed their efforts to settle the region.

Above: The rocky coastline of the Gulf of Saint Lawrence around Cape Breton. Cartier's report of a great inland sea encouraged further exploration by a French crown eager to rival European discoveries elsewhere in the New World.

AFRICA
DIOGO HOMEM

During the great Age of Discovery, cartographers' skills were in demand and the knowledge they acquired while working for one royal employer made them even more valuable to rival rulers. Consequently a chartmaker like Diogo Homem could leave his native land, confident that his services would be appreciated elsewhere.

Homem was the son of Lopo Homem, master of sea charts to the king of Portugal. Diogo worked with his father but little is known of his life until, in the 1550s, he fled Portugal after killing a man in a fight. He settled in England, where he drew an atlas which included the maps of East Africa and the East Indies illustrated in this book. The atlas was presented to Queen Mary in 1558. Homem soon moved to Venice, and in 1569–1571 published a chart of the Mediterranean and northern Europe which has claims to be the first printed sea chart. Homem was one of the leading chartmakers of his time; the survival of his work is all the more important because so many maps from the period were jealously guarded by the Crown and eventually destroyed. The map of Africa and the Indian Ocean is drawn in portolan-chart style but, though informative, it is essentially a presentation piece, a feast for royal eyes.

ISLAMIC TRADING PARTNERS
An Islamic flag, with golden crescents, marks the eastern Mediterranean, controlled by the Ottoman Empire.

EXOTIC PLACES
The Portuguese learned little of the interior of countries where they traded and founded bases. Exotic scenes were included to delight the eye—and fill blank spaces.

MYTHICAL MONARCH
The legendary Christian king Prester John appears, although Portuguese agents who reached Ethiopia in the fifteenth century failed to find him.

FAMILIAR COASTLINE
By contrast with the pictorial treatment of interiors, the sea coasts are carefully drawn and filled with place-names, reflecting the regular voyages to India and beyond made by the Portuguese.

FICTIONAL TOPOGRAPHY
Imaginary rivers, lakes and mountains—including a distant source for the Nile—fill up the African interior. In reality the Portuguese were content to build forts on the coast and trade from them.

A ROUTE TO INDIA

Homem's maps of Africa and the Indian Ocean describe a region that had been penetrated and exploited by his compatriots. Driven by lust for gold, slaves, and spices, Portuguese epic voyages were initiated by Prince Henry the Navigator and sustained by strong-minded monarchs.

Henry, the third son of King John I, spent almost no time at sea and so hardly seems to deserve his sobriquet "the Navigator." His place in the history of exploration is secure, however, thanks to the geographical research center he established at Sagres and the voyages he organized down the African coast. From 1434 expeditions were sent south, beyond the three known island groups off the African coast, toward regions where, it was feared, mariners would be baked black by the tropical sun or boiled alive in solar-heated seas. Henry hoped to locate the source of the trans-Saharan gold trade, which was known to be in some distant part of West Africa, and also to contact the legendary Christian ruler Prester John. Neither quest produced satisfactory results, but in 1441 Africa did begin to yield substantial profits, wrung from the infamous slave trade.

Headland by headland, the Portuguese advanced, reaching Cape Verde (Dakar, Senegal) by 1448. From there the coastline ran southeast and then east, raising hopes that Africa had finally been rounded and that a route to the East lay open. Henry died in 1460, before the Portuguese reached what proved to be the Gulf of Guinea. Its almost 1,250 miles (2,000 km) of coastline never turned north, as they had expected, but eventually swung around in the opposite direction, stretching away southward without any visible end.

Sailing around Africa

Fears of being baked or boiled had proved unfounded. And since experience had shown that the ships, weapons, and forts of the Portuguese were more than sufficient to enable them to dominate the peoples they encountered, the possibility of finding a sea route to the East seemed increasingly likely. After the strong-minded John II inherited the throne in 1481, state-sponsored expeditions produced impressive results. The voyages made by Diogo Cão in 1482–1485 took him some 1,850 miles (3,000 km) beyond the equator, probably as far as Walvis Bay (Namibia). Then Bartholomeu Dias, despatched in 1487 with three ships, rounded the Cape of Good Hope and gazed on the Indian Ocean.

All the Portuguese expeditions were conducted in great secrecy; any maps they used have disappeared and many details of the voyages remain uncertain. It is not clear, therefore, whether Dias was blown into the South Atlantic by storms or already knew that he must sail out into the ocean in order to pick up the

Left: Henry the Navigator, holding a model of a caravel, looks out over the Tagus River in this modern monument to the Age of Discovery in the Portuguese capital, Lisbon.

westerlies to be driven round the Cape. However it happened, Dias's ships were blown round to Mossel Bay and went on to the mouth of the Great Fish River, northeast of present-day Port Elizabeth, before mutinous crews forced him to turn back.

Dias had finally proved Ptolemy wrong: Africa was not joined to the supposed southern continent, and so a sea route from Europe to the East did exist. Even so, it was not until July 1497 that an expedition left Lisbon under orders from a new king, Manuel. Led by thirty-seven-year-old Vasco da Gama, it did not hug the coast but sailed to the Cape Verde Islands and then swung out into the ocean. The winds only carried the ships to a point just north of the Cape on the west coast, and they had to work their way around before sailing up the eastern side of the continent. There they found cities and long-established Arab traders, not easily dominated and jealous of their monopoly. After some

vicissitudes, da Gama took advantage of the monsoon winds that in spring and summer carried ships across to the west coast of India, and in fall carried them back again. He reached the great city of Calicut, where he acquired a cargo of pepper and cinnamon, and returned to Portugal in September 1499.

In 1500 a new expedition was sent out under Pedro Alvares Cabral. Since da Gama had not succeeded in rounding the Cape in a single sweep, Cabral swung out into the Atlantic in a wider arc—so wide, in fact, that he reached Brazil, effectively beginning the Portuguese involvement with that land. From a Portuguese point of view, the East remained more important. Forts and trading stations were established in East Africa, India, and the Malaysian archipelago, and when their vessels reached the fabled Spice Islands (the Moluccas), the Portuguese were able to reap the full rewards of their daring and persistence.

Above: The end of the land: the Cape of Good Hope at the southern tip of Africa. Bartholomeu Dias's successful—though possibly inadvertent—rounding of the Cape opened a new era of European sea exploration, trade, and colonialism.

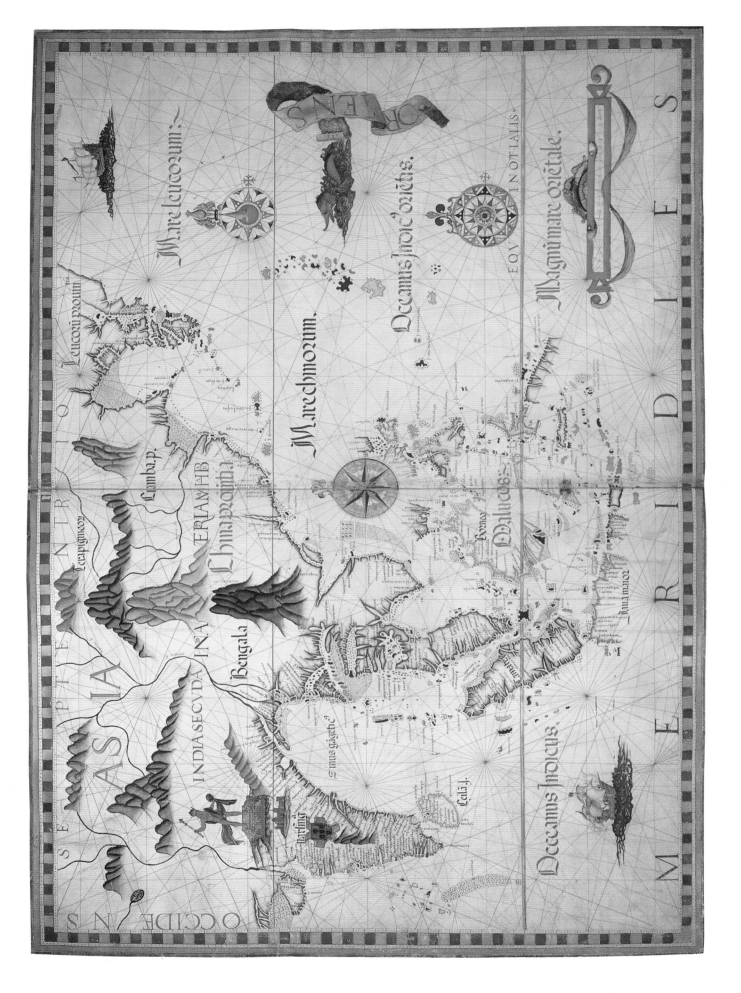

ASIA
DIOGO HOMEM

Christopher Columbus sailed west but failed to find "the Indies," the source of the spices that Europe craved. The Portuguese sailed east and, having located them, became immensely wealthy. With geographical information at a premium, the chart made by the fugitive Portuguese cartographer Diogo Homem represented a form of classified information that was greatly valued by his new employers.

Homem's chart was part of the manuscript atlas that he presented to Queen Mary of England in 1558. Mary was married to King Philip II of Spain, and Homem may well have been attracted to London by the connection with Spain. At that time, Spain was Portugal's great rival and likely to be the most generous paymaster for a Portuguese mapmaker on the run from a murder charge in his native land. Ironically, Mary died soon after the presentation, the Anglo-Spanish alliance dissolved, and the atlas remained in a Protestant England that would later prey upon the shipping and colonies of Spain. The economic importance of the Spice Islands is brought home by their central place in Homem's work. Though designed to please monarch and court rather than be taken to sea, the chart is still in most respects a typical portolan chart, no different in kind from the Carte Pisane almost 300 years earlier (page 36).

OUTSIZE HARBORS
Here and elsewhere, the traditional style of the portolan chart persists, with features such as harbors greatly exaggerated and place names virtually restricted to areas along the coast.

ABSENT JAPAN
Korea is known, if not very well, since it appears as more archipelago than mainland. Homem has wisely omitted Japan, of which no accurate reports reached the West until the 1590s.

WEEDY
An amiable-looking, weed-wrapped sea creature enlivens the Oriens (East) label and seems to be pointing to the Marianas, discovered in 1521 by Ferdinand Magellan, who called them the Ladrones ("Thieves' Islands").

SPICE ISLANDS
The Moluccas (Malucos), or Spice Islands, are literally writ large on account of the wealth generated by their cloves, mace, nutmegs, sandalwood, and other valuable resources.

SECRECY, SPYING, AND SPICES

Diogo Homem was far from unique in leaving his native Portugal to serve its foreign rivals. In an age of great discoveries, a mapmaker with inside information and sufficient expertise might literally be worth the wealth of the Indies to one of the maritime powers. Maps became closely guarded state secrets, but, as always, they were vulnerable to bribery, leaks, and defections.

Initially Portugal was the kingdom with the most important cartographic secrets to lose. After the discovery of the Americas, their economic value was not obvious. By contrast, gold and slaves lured Portuguese caravels down the coast of Africa, and once Vasco da Gama had reached India, there were vast profits to be made from trade with the East. As early as 1479, Portugal claimed West African waters as its exclusive zone of navigation, and decreed that any foreign crews captured there would be thrown overboard to drown. Proposals to omit new discoveries altogether from charts and maps were hardly feasible, but navigators and cartographers were sworn to secrecy and threatened with execution for breaches of security.

Despite the risks involved, in 1502 Alberto Cantino managed to smuggle a world map out of Lisbon. It contained up-to-date information about the Portuguese and Spanish discoveries in both the Old and the New Worlds. Many place names are said to have been supplied by Amerigo Vespucci, but the identity of the mapmaker is one secret that has never been revealed. Cantino was the agent of a leading Italian Renaissance prince and patron, Ercole d'Este, duke of Ferrara. It is not clear whether Ercole was merely curious, or worried about the economic effects of the new routes.

The Venetians were certainly concerned, especially after an efficient intelligence operation revealed the extent of the Portuguese success. About 1505 a Genoese, Nicolò Caveri, produced a world chart that seems to have been based on (or came from the same source as) Cantino's, with some new information about Portuguese-ruled Brazil. It was thanks largely to such leaked information reaching Italy that details of

D. VASCO DA GAMA. VI.

Above: In 1498 the Portuguese mariner Vasco da Gama opened the sea route to the East when he sailed around Africa into the Indian Ocean. Later the Portuguese reached the Spice Islands.

the discoveries were disseminated north of the Alps, benefiting cartographers such as Waldseemüller.

The East held many prizes for Europeans, but the greatest of all were the products of the Spice Islands, or Moluccas (now part of eastern Indonesia), notably cloves, mace, nutmeg, and sandalwood. The Portuguese found their way to the islands in 1511, but their monopoly was by no means safe from their Spanish rivals. In 1494, after the pope had decreed that all newly discovered lands should be divided between Spain and Portugal, the Treaty of Tordesillas divided the world by a north–south line 370 leagues to the west of the Azores: everything to the west was Spain's; everything to the east (Brazil, Africa, India) was Portugal's. There was no chance that other nations would accept these arrangements, but for a long time the Spanish and Portuguese would be the only contenders in the Southern Hemisphere.

Who owns the Spice Islands?
The main object of their contention would be the Spice Islands. Which half of the Spanish–Portuguese world did the islands lie in? The answer depended on the distance between the New World and the islands. Portuguese mapmakers in Spanish pay, Diego Ribeiro and Jorge Reinel, made maps to help their fellow Portuguese, Ferdinand Magellan, make a swift passage from Cape Horn to the Spice Islands.

Although the Spanish expedition commanded by Magellan did reach the islands and sail around the globe in 1519–1522, the "swift" passage across an unexpectedly vast Pacific lasted for months. Magellan's expedition finally discredited Ptolemy's elongated version of Asia and diminished the attractions of a westward, transatlantic route to the East.

With so much at stake, a cloak-and-dagger atmosphere prevailed throughout this period. The Spanish appear to have been highly successful in luring away Portuguese cartographers or buying information from them. Reinel was brought back to Portugal by his more distinguished father, Pedro. According to an ambassadorial report, however, Pedro, during his stay in Seville, updated a Spanish chart and even drew the Spanish–Portuguese line on a map so that the Spice Islands fell in the Spanish hemisphere. Since Pedro remained in favor on his return to Portugal, either the report was wrong or he was feeding the Spanish misleading information—a murky situation typical of the world of espionage.

Other reports suggest that the equally distinguished Lopo Homem, father of Diogo, might be bought. Even much later, after Diogo's flight, cartographers were sufficiently important for the king to promise that the murder charges against Diogo would be dropped if he returned to Portugal. Unconvinced, or better paid abroad, Diogo stayed away. Nevertheless, Portugal held on to its eastern empire until the end of the century, when a new maritime power, the Dutch, seized control of its much-coveted Spice Islands.

Above: Garlic and spices on sale in a modern market on Lombok, one of Indonesia's Lesser Sunda Islands. Spices were a luxury ingredient in the early modern diet, essential for livening up the taste of food that was at best monotonous, at worst rotten.

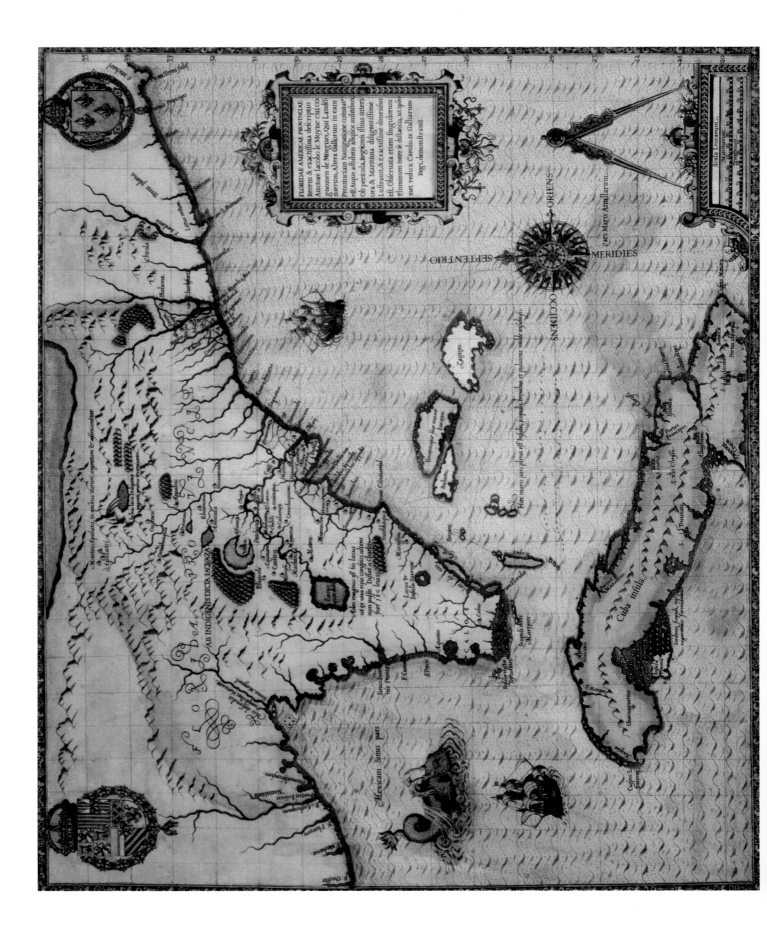

FLORIDA
JACQUES LE MOYNE

Despite their successful penetration of the Caribbean and Mexico, Europeans found it difficult to establish permanent settlements on the North American mainland. Success was eventually achieved in Florida, but only after a number of failures—one of which lies behind the map drawn by Jaques Le Moyne.

In 1564 the artist Jacques Le Moyne accompanied a French expedition led by René de Laudonnière to found a settlement on the coast of Florida. The party was massacred the following year by hostile Spanish forces. Le Moyne was one of those who escaped, but his map of the region and drawings of Native American life were not published until 1591. The originals have disappeared, but they are preserved in the fine engravings made by Théodore de Bry, along with Le Moyne's own notes and his narrative of the expedition. De Bry worked on the English version of Waghenaer's navigation charts (page 116), and his contacts with Richard Hakluyt, the chronicler of English voyages, gave him the idea for his own *Grands Voyages* (to America) and *Petits Voyages* (to the East). These collections of narratives with maps and illustrations included Le Moyne's experiences. De Brys' engravings were widely copied, and by providing materials for border scenes, they played a large part in creating the fashion for elaborately decorated maps.

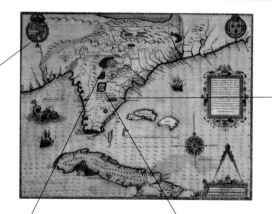

FORT CAROLINE
"Carolina" is the name used on the map for Fort Caroline, the site of the French settlement on the St. John's River where Le Moyne lived.

COAT OF ARMS
The coat of arms of the Spanish Hapsburgs. In the opposite corner, the royal French coat of arms, with its three fleurs de lys, represents the other claimant to Florida.

MOUNTAIN WEALTH
The Appalachian Mountains "in which are found gold, silver, and copper." The lake below, according to the caption, posssesses "grains of native silver."

MIGHTY LAKE
Lake Okeechobee is fairly accurately drawn from observation or report. It is described as so great that "from one bank the other cannot be seen." Cape Canaveral is marked as lying to the southeast, although in reality it is much farther north.

EXPLORERS AND SETTLERS

In the southern New World, Europeans were motivated by lust for gold and the illusory quest for a passage to the East. In North America, where glittering prizes were not on offer, founding settlements required a different attitude, including a willingness to face difficulties for relatively humdrum rewards.

The Spanish and Portuguese monopolized the occupation and exploitation of Central and South America, and they claimed sovereignty over the entire New World. States such as France and England disputed the claim, but the only openings for them were in the northern continent. There, the absence of gold-rich cities proved a disappointment, and the real opportunities were grasped only slowly.

French efforts began in the far north, where Bretons were probably fishing the waters off Newfoundland by 1500. Exploration of the Gulf of St. Lawrence led to a short-lived attempt to settle Sable Island as early as 1518. In 1524 Giovanni da Verrazano, a Florentine in the service of France, explored the east coast, and between 1535 and 1541 Jacques Cartier made his three pioneering voyages down the St. Lawrence River. On the third voyage he tried to establish a colony at Québec, but its failure ended French activity until the end of the century.

English interest in the area was also sporadic. In 1497 another Italian in foreign service, John Cabot (Giovanni Caboto), reached Cape Breton Island and Newfoundland. A second voyage in 1498 convinced him that the riches of Asia did not lie in the vicinity, and the English lost interest in North America. It revived under Queen Elizabeth, taking the form of

Above: The first French arrivals receive a friendly welcome from Florida's native inhabitants in this sixteenth-century illustration based on a sketch by Jacques Le Moyne.

slave-trading, and then of piratical attacks on the Spanish, notably by Sir Francis Drake during his voyage around the world from 1577 to 1580. The principal English activity in the 1570s and 1580s was the quest for a northwest passage to the East and its riches, sought in vain by Martin Frobisher, Sir Humphrey Gilbert, and John Davis.

The first English colony

The founding of an English colony, to be named Virginia after England's Virgin Queen, Elizabeth, was advocated by Gilbert, and after his death in 1583 by Sir Walter Raleigh. In 1584 a settlement was established on Roanoke Island (now North Carolina), but the colonists quickly returned to England. A second party, sent out in 1587, fared even worse: neglected during Armada year (1588), it had vanished without trace by the time relief arrived. John White, artist and mapmaker, was one of the first settlers and was the governor of the second Roanoke colony. He was absent in London when the colony disappeared, so his beautiful watercolor maps and paintings survived, many of them published by de Bry in his *Grands Voyages*. Despite the early disasters, the English colonizing impulse persisted, finally bearing fruit in the early seventeenth century.

Meanwhile, the Spanish, already possessors of an enormous empire, were also exploring and expanding into North America. Many of their feats were quite spectacular. In 1528 an expedition led by Pánfilo de Narváez was wrecked on the Texas coast. Most of its members died, but Alvar Nuñez Cabeza de Vaca and three others made their way right across Texas and returned to Mexico.

In 1539 a Franciscan friar, Marcos de Niza, reached the pueblos (native towns) of New Mexico. Inflamed by tales of the Seven Cities of Gold, Francisco Vásquez de Coronado and Hernán de Alarçon followed in 1540–1542, going on to Colorado, northern Texas, and eastern Kansas before turning back.

During the same period, Hernán de Soto made an equally extraordinary journey from Florida. He discovered the Mississippi River, traveled as far as Oklahoma, and then followed the Arkansas River back to the Mississippi.

Settling was another matter. Ponce de León discovered Florida in 1513 but failed to colonize it. Later attempts were so unsuccessful that by 1560 they had been abandoned. But when the French settled at Fort Caroline in 1564, Pedro Menéndez de Avilés was sent to assert Spanish rights. In 1565 he not only destroyed Fort Caroline but also founded the colony of St. Augustine. It was a modest success, and the Spanish became the first Europeans to establish a permanent settlement in the future United States.

Above: Juan Ponce de León, Spanish governor of Puerto Rico, discovered Florida by accident while searching for an island said to be where the Fountain of Youth was located.

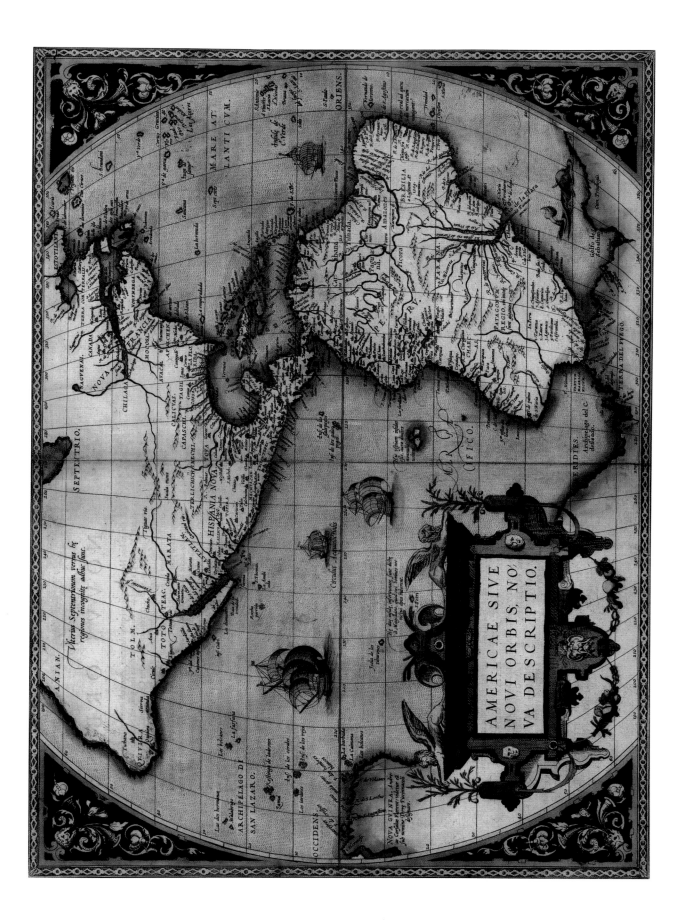

THE NEW WORLD
ABRAHAM ORTELIUS

This map of the New World is a fine example of the sixteenth-century engraver's art.
It formed part of the pioneering atlas issued in 1570 by the Antwerp mapmaker Abraham
Ortelius. Though unusually conscientious in correcting his maps in later editions,
Ortelius was inevitably misled by contemporary exaggerations and fantasies.

Ortelius's first map of the New World was closely based on the researches of Gerardus Mercator and other contemporary sources; consequently its errors and limitations were not personal but reflected the state of European knowledge. The map implicitly celebrates the feats of European explorers, settlers, and traders by picturing the oceans as mastered and the continents filled with named river systems, mountains, and other features. Though exaggerated, the impression of intense activity was not unjustified, and the arrival of new information enabled Ortelius to update his work regularly; in particular, the large bulge visible here on the western coastline of South America was eliminated, and the Solomon Islands were correctly placed if much over-enlarged. However, mythical islands were still to be found in the Atlantic, and Ortelius and later cartographers grossly overestimated the size of far-northern America and believed in the existence of a vast, undiscovered southern continent.

FIRST GUESS
The island of New Guinea, discovered about fifty years earlier, appears as part of the southern continent. Ortelius corrected his mistake nineteen years later, when he issued the first printed map of the Pacific.

OPTIMISM
These islands are shown even though, in almost a century of Atlantic crossings, sailors had found nothing between the Azores and Bermuda. "Sept cites" optimistically identifies the legendary Seven Cities of Gold.

PATAGONIA
Ortelius notes that the Patagonians are giants—but they are not, apparently, city-dwellers. A list of Peruvian cities occupies the empty land, keyed to numbers higher up on the name-crowded west coast.

SOUTHLAND
Tierra del Fuego is shown as part of the mythical southern continent, although it is in fact an island. The coastline stretching away on either side of it is purely imaginary.

THE FIRST ATLAS

Ortelius's map of the Americas was part of a collection that had been organized on new principles. Systematic in coverage, uniform in size, and bound together as a volume, the maps of the *Theatrum Orbis Terrarum* ("Theater of the World") constituted the first atlas. Published in 1570 and reissued many times, it became the cartographic bestseller of the century.

The idea for the *Theatrum* is said to have originated with a Dutch merchant named Hooftman who collected maps. Irritated by the need to roll and unroll his treasures, he employed Ortelius to bind together a bundle of similar-sized maps so that they could be consulted more conveniently, in book form. Ortelius realized that just such a book of uniform maps, describing the known world, would have a strong public appeal, and the concept of the atlas was born.

The execution of the project was favored by the fact that Ortelius was as much a businessman as a scholar. Born in 1527, he spent his working life in his native Antwerp in the Low Countries, then at the height of its commercial greatness. He started a business as a dealer in books, maps, and antiquities, also advertising his services as a "map colorist." He evidently prospered and was able to travel for business and pleasure in Germany, France, and Italy, where he built up a network of contacts among scholars and cartographers. He also spent some time in England, possibly as a

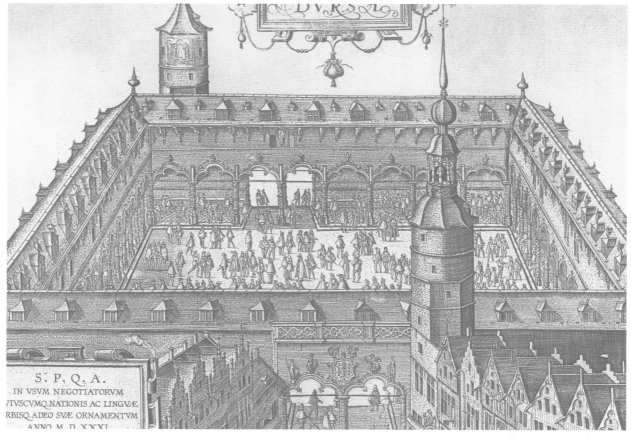

Above: For an entrepreneur launching a new business, few cities were as promising as Antwerp. The city was one of Europe's richest, with merchants and financiers growing wealthy on the thriving stock exchange, or Bourse, pictured here in around 1530.

Above: The success of the *Theatrum* prompted Philip II of Spain, the Hapsburg ruler of the Low Countries, to appoint Ortelius as his court geographer.

religious refugee. During his stay there he met Richard Hakluyt, author of famous narratives of maritime exploration, and encouraged William Camden to complete *Britannia*, his pioneering study of the island's antiquities.

A prolific mapmaker

An ardent collector and a friend of Gerardus Mercator, Ortelius was well placed to produce maps on his own account. His first, published in 1564, was an eight-sheet world map in a distinctive heart-shaped projection. It was followed by maps of Egypt (1565) and Asia (1567). Work on the *Theatrum* must have already been under way at the time, since its seventy maps must have taken years to prepare as information was gathered and collated from sources all over Europe. We know that the contents were based on the work of eighty-seven earlier authorities, thanks to Ortelius's exceptional candor in acknowledging his cartographic debts. Earlier maps had to be updated and redrawn to a uniform size. Then they were engraved, mostly by the gifted Frans Hogenberg, also remembered for his work on the first great collection of city maps (page 108).

The first edition of the *Theatrum* contained fifty-six maps of Europe, ten of Asia and Africa, and one of each of the four continents. This certainly made it an atlas, although the word would not become current until after 1585, when Ortelius's friend Mercator used it as the title of his own published collection (Atlas, a popular image on maps, was a mythical Titan who held up the heavens).

The *Theatrum* enjoyed an immediate success, reprinting within a few months. It was frequently revised and expanded—for example, taking in the first separate maps of China (1584) and Japan (1595). In 1575 Ortelius was appointed geographer to King Philip II of Spain, the ruler of the Low Countries. He went on to publish historical maps, the *Parergon Theatri*, from 1579, which were also incorporated in editions of the *Theatrum*. By the time of his death in 1598, Ortelius's atlas consisted of 119 contemporary and thirty-six historical maps. The *Theatrum* continued to be published until 1612, appearing in Latin and in the principal European languages. A 1606 English edition contained no fewer than 161 plates.

The popularity of the *Theatrum* soon brought competitors into the field, and before the end of the sixteenth century, atlases were published by Gerard de Jode and Mercator. Neither achieved a commercial success to compare with that of Ortelius, but their efforts helped to establish the atlas as a prestigious and useful form in which to possess a set of maps.

Above: An investment for the future: armed soldiers guard merchants setting sail from Antwerp harbor. Continued prosperity depended on protecting the city's lucrative trade.

MUSCOVY
JENKINSON AND ORTELIUS

The English took little part in the great age of discovery and expansion that began in the late fifteenth century. Then there was a burst of activity under Queen Elizabeth, when famous voyages were made to America and beyond. There were equally daring ventures into Arctic waters, and these led to contacts with the distant state of Muscovy, or Russia.

The earliest maps of Muscovy, published in the 1540s, were drawn by western Europeans on the basis of information supplied by Russian envoys or émigrés. The first map based on first-hand knowledge was made by an English merchant adventurer named Antony Jenkinson. English involvement with Muscovy began in 1553, when Chancellor and Willoughby's expedition tried to find a northeast passage to the East through the Arctic seas. Willoughby perished in the White Sea, but

Chancellor managed to make his way down the Dvina to Moscow. The contact was promising enough for English merchants to form a Muscovy Company and send out Jenkinson as their representative in 1557. He successfully established commercial relations with the czar, Ivan IV ("the Terrible"), and then traveled down the Volga to the Caspian Sea and on to Bukhara. In 1562 he published a map of Muscovy, but it has not survived except as a plate in Ortelius's atlas of 1570, where Jenkinson's authorship is clearly acknowledged.

RIVER DVINA
The first English contacts with Russia were not made via the Baltic but by sailing over the top of Scandinavia into the White Sea. From there the River Dvina led to the interior of Russia.

ASTRAKHAN
Antony Jenkinson sailed down the Volga to Astrakhan and the mouth of the Caspian. Here, conscious of being the first Englishman to reach this inland sea, he set up a flag carrying the cross of Saint George, the patron saint of England.

STRANGE CUSTOMS
The funeral customs of the Kirghiz are claimed to involve suspending the dead from trees and ritually sprinkling the living with blood and milk. Other scenes, such as the ones above this detail, serve to illustrate the idolatries and strange customs attributed to native peoples.

PRINCIPALITY
Lying at the heart of a great river system, Moscow grew from a small principality to the sizable state shown on Ortelius's map. Within years it would expand still farther to the east.

THE RISE OF RUSSIA

Antony Jenkinson's map of Muscovy shows a strong Russian state centered on Moscow. Its existence represented a triumph for the policies pursued by a series of able princes. The emergence of czarist Russia as a great power was by no means certain, however, and it was achieved only after 150 years of turmoil.

A Slav people—later called the Russians—settled the steppes north of the Black Sea centuries before the foundation of the first Russian state. That took place in the ninth century, when the Slavs fell under the sway of Viking invaders, the Varangians, or Rus, who made Kiev their capital.

The Varangians rapidly became Russian in language and outlook, but their Viking-style raiding brought them into contact with the Byzantine Empire and its Greek and Christian culture. As a consequence Kiev Rus was converted to Orthodox Christianity and adopted the Greek alphabet. Russia became part of Christendom but, like Byzantium itself, was different in some important respects from the Latin and Catholic West—a difference that would have an important influence on the Russians' later attitudes to their European neighbors.

Flight and settlement

Kiev Rus flourished for a time in the eleventh century, but later it fragmented and a nomadic Turkish people, the Cumans, overran much of its territory. Fleeing Russian peasants settled vast areas in the northeast, founding the independent republic of Novgorod and beginning the shift of the Russian center of gravity to the east, away from Europe.

Between 1219 and 1242 all the Russian principalities except Novgorod were overrun by the Tartars (as Ortelius and many others called the

Above: Ivan the Terrible as imagined by a nineteenth-century artist, mourning beside the body of his son, whom he murdered in a fit of rage. Ivan's reign exemplified the strengths and often appalling drawbacks of autocracy.

Mongol Golden Horde that then dominated the region). For the next two centuries the Russians recognized the Tartars as their overlords and paid tribute to them. Meanwhile Catholic powers—the Teutonic Knights, Sweden, and Lithuania—expanded eastward at Russian expense. Among their conquests was Ukraine, including historic Kiev.

Of the remaining Russian principalities, it was Vladimir that took the lead at first, thanks to its prince, Alexander Nevsky, who won famous victories over the Swedes and the Teutonic Knights. As Vladimir went into a decline, however, Moscow gradually came to the fore. The city dated only from the twelfth century, but its easy access to the Volga and other rivers was a strategic and commercial advantage. It was also fortunate in having an almost unbroken series of able rulers, shrewd enough to act as tribute collectors for their Tartar overlords. Thus Moscow became the nucleus of a principality that slowly consumed its rivals. Finally, in the 1470s and 1480s Ivan III subdued Novgorod, renounced his allegiance to the Golden Horde, and effectively created a Russian national state.

Ivan the Terrible

The Muscovite ruler whom Chancellor and Jenkinson met was Ivan IV. "Ivan the Terrible" was the first prince to be crowned as czar, the Russian version of "Caesar." His autocratic policy led to conflicts with the great nobles, or boyars, who were forced to submit, and the growing weakness of the Tartars enabled Ivan to drive them from Kazan and Astrakhan. Later in his reign, however, Ivan's homicidal tendencies became increasingly apparent, and a reign of terror (the victims including his own son) caused chaos. One result was that a painfully acquired stretch of Baltic coast was lost again, and Muscovy remained a landlocked power except for the entrance to the White Sea, where Ivan built the port of Archangel.

After Ivan's death in 1584, the boyars regained much of their power, the ruling line died out, and pretenders, peasant revolts, and foreign interventions brought Russia close to disintegration. This "Time of Troubles" came to an end when the Russians rallied to drive out Polish invaders and Michael Romanov was elected as czar. His Romanov dynasty ruled Russia down to the 1917 Revolution.

In the seventeenth century a tacit understanding stabilized the political and social situation. The czars were accepted as autocrats; in return, the nobility became masters of a peasantry reduced to serfdom. Consequently, the reforming autocrat Peter the Great was able to impose a ruthless westernization on backward Russia. In 1709 his decisive victory over the Swedes gave Russia access to the Baltic and, in newly built St. Petersburg, a "window on the West." Russia had emerged as a formidable state, and would from this time forward have to be reckoned with in the European balance of power.

Above: Saint Basil's Cathedral, one of Moscow's most famous sights, was built by Ivan IV between 1554 and 1560 to celebrate the defeat of the Tartars of Kazan and Astrakhan.

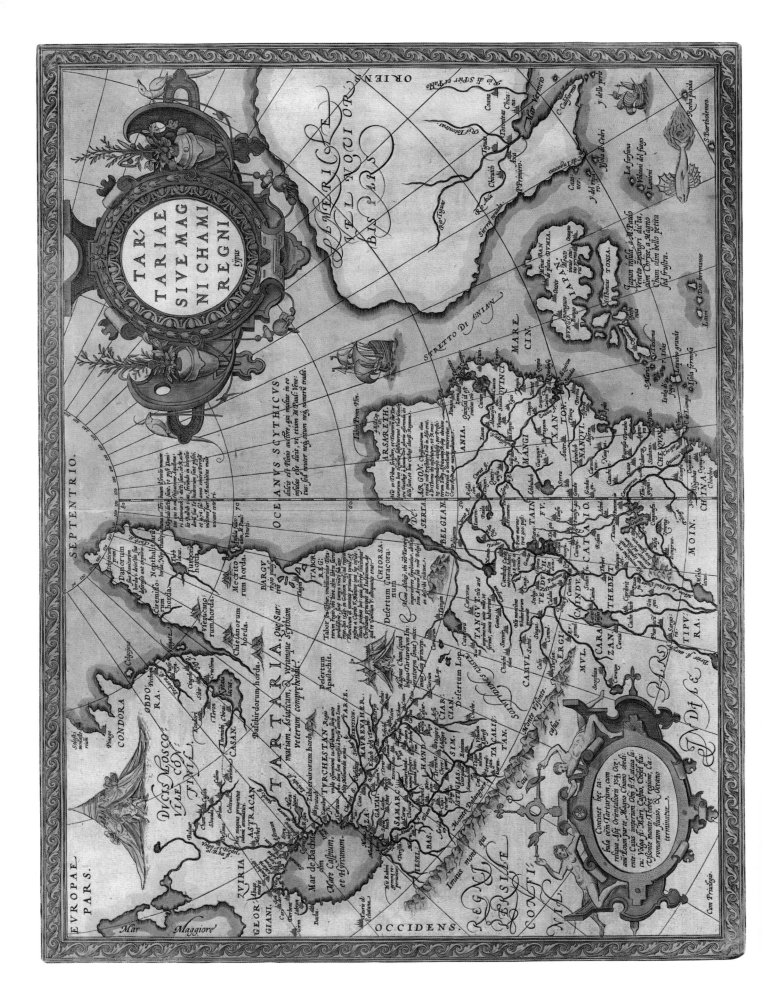

TARTARY
ABRAHAM ORTELIUS

Northern and eastern Asia were still little known in Ortelius's day, though reports from Jesuit missionaries had begun to bring the shape of China into focus. Here Ortelius boldly tries to map the dominions of the Tartars, or Mongols, in 1570. The result is like a companion to Marco Polo's *Travels*, representing Asia as it had been two centuries earlier.

This was a much copied map, perhaps because of its visual and exotic appeal. Much of it seems based on the accounts of Marco Polo and other travelers who crossed the Mongol Empire when it stretched from Persia and Russia to China. After the death in 1294 of Kublai Khan, the empire broke up into rival khanates. The Mongol emperors who ruled China were expelled in 1368, but Ortelius shows it as still part of the dominions of the "Great Cham" (Great Khan). West of Tartary lay the territories of the dukes of Moscow, who had long acknowledged the Mongols as their overlords. By 1570, however, Duke Ivan IV ("Ivan the Terrible") had become the Russian czar and had begun to expand eastward at the expense of the Mongols, capturing Kazan ("Casan") in 1552. Ortelius's map shows some knowledge of Central Asia and China, but much of Asia (notably Siberia) is compressed and distorted. Optimistically, the Pacific is shown as presenting no great barrier to travel, with Japan conveniently close to California.

A COVER-UP
The splendid cartouche, as on some other European maps of the time, serves the secondary purpose of concealing the mapmaker's uncertainties. Here it begs the question of whether a northwest passage to Asia lies behind it.

ANIAN STRAIT
North America and Asia are separated by the fabled Anian Strait, while Japan provides a useful stopover between an abbreviated California and the coast of China.

KARAKORUM
Tents mark the site of the Mongol capital, Karakorum. Even when they were masters of China, the Mongols retired to their own land each summer to resume the hardy lifestyle that had made them invincible.

MISSING SEA
Though familiar Central Asian names such as Samarkand appear on the map, the Aral Sea is missing and the great rivers that flow into it are wrongly linked to the Caspian.

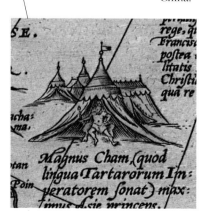

THE MONGOLS

Published in 1570, Ortelius's map of Tartary evoked the earlier glories of the Tartars, or Mongols, whose Great Khan had ruled Asia from Muscovy to China. By Ortelius's day the empire had broken up into rival khanates. Some of these proved extremely durable, but the forward impetus of the Mongols had clearly spent itself by then.

The Mongols were the most formidable of all the pastoral nomads who attacked and sometimes overran settled agricultural societies. Constantly on the move, such nomads became expert horsemen, inured to hardship, for whom the rich cities of Asia and the Mediterranean represented particularly tempting prizes. In the seventh century the irruption of nomadic Muslim tribes from Arabia changed history. Most often, however, the onslaught originated from somewhere in the vast steppes that stretched from Hungary to Manchuria.

The Mongol homeland was a bleak plateau to the north of modern Mongolia. Tribal divisions restricted the Mongols' impact until 1206, when a chief's son, Temujin, who had more than overcome the hazards of a fatherless boyhood, was proclaimed Genghis (or Chingis) Khan, "Lord of the Earth."

Genghis rapidly demonstrated that he was a leader of genius, binding Mongol society together with a legal code and creating a highly disciplined army, able to take full advantage of its great mobility. Over the next twenty years, Genghis's armies forced their way across the steppes, enrolling other nomad peoples under their banners. They invaded China and captured Beijing, defeated the Russians, and overran much of the Near East and Persia, capturing the famous cities of Bukhara and Samarkand.

A vast empire

Genghis reached the banks of the Indus in 1224, when he returned to Mongolia to deal with a revolt. He died there in 1227, having divided his empire between his sons. One of them, Ogadai, was in supreme authority as Great Khan. Under Ogadai and his successor, Mangu, the Mongols captured Baghdad, overthrew

Above: Mongol justice in action: in this contemporary illustration Genghis Khan looks on as a prisoner is flogged. The Mongols' vast empire was built on terrible violence.

the Abbasid dynasty, and conquered north China and most of Russia. Europe was spared only because of an untimely interregnum following the death of Ogadai.

Mongol policy was based on spreading terror by massacre and destruction; some regions took centuries to recover from the thoroughness with which they were laid waste. The positive side of Mongol rule was the pacification of a vast area through which travelers could pass in safety. The

"Pax Mongolica" was at its height under the third Great Khan, Kublai, whom Marco Polo the Venetian served in the thirteenth century.

A slow decline

By the late 1270s, Kublai had completed the conquest of China, and the Mongol Empire had reached its greatest extent. Even under Kublai, the various khanates ruled by members of his family were virtually independent, but he retained nominal supremacy and the peace held. After his death in 1294, the khanates rapidly became rivals, waging war on one another, and East–West contact was again disrupted.

A slow decline followed. The Mongol khanate of Persia broke up in 1334, and the Mongol Yuan dynasty was expelled from China in 1368. The khanate of the Golden Horde continued to occupy much of what is now European Russia, but like the other Asian khanates, it gradually shrank and fragmented.

The Mongol impulse was briefly renewed in the late fourteenth century, when the bloodthirsty Timur (Tamerlane) carved out an empire from the Black Sea to India. Although it collapsed after his death in 1405, in 1526 Babur, a descendant of Timur, swept down from Afghanistan into India and founded the long-lived Mughal (Mongol) dynasty.

Above: A stone tortoise and abandoned Buddhist stupas mark the site of Karakorum, Genghis Khan's capital in the bleak grasslands of Mongolia. The city had been destroyed by Chinese armies in 1388, almost two centuries before Ortelius showed it on his map.

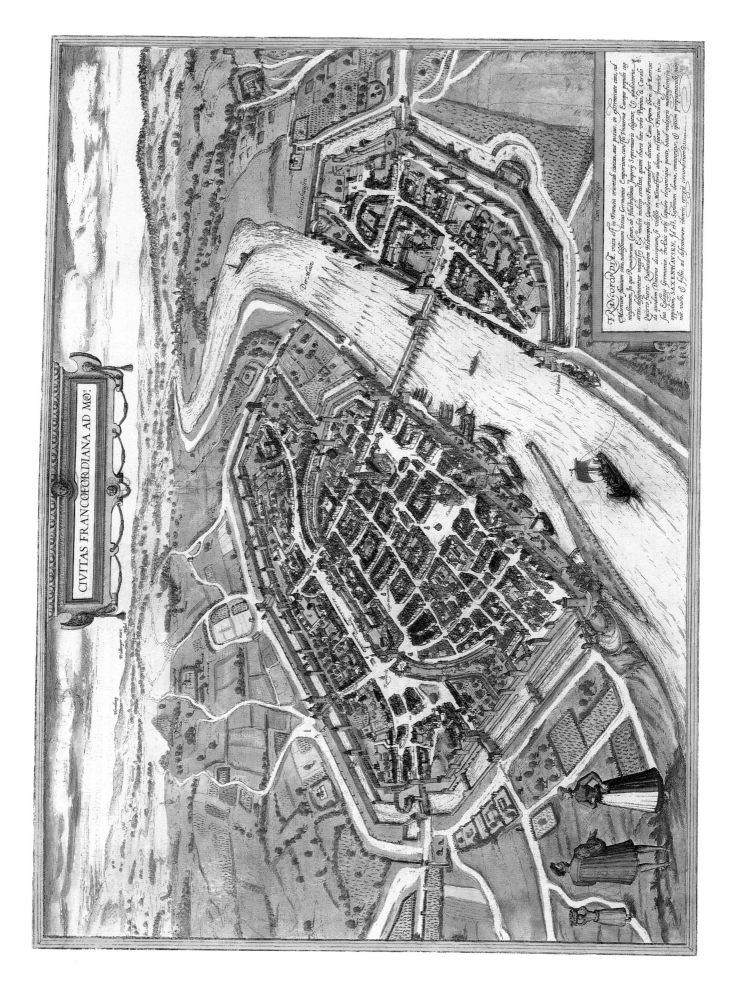

CIVITAS FRANCOFORDIANA AD MŒ.

FRANKFURT-ON-MAIN
BRAUN AND HOGENBERG

The German city of Frankfurt-on-Main was a leading commercial and ceremonial
center in the Holy Roman Empire. It appeared in the first of six pioneering
volumes of town plans, *Civitates Orbis Terrarum*, issued by Georg Braun and
Frans Hogenberg between 1572 and 1618.

rankfurt played an important role in German
history from the time of Charlemagne in the
eighth century. The Holy Roman emperors were
elected there from 1356, and Frankfurt became an
Imperial Free City in 1372. From 1562 the coronation
of the Holy Roman emperor took place in Frankfurt's
great cathedral of St. Bartholomaüs.

Braun and Hogenberg's map of Frankfurt, based
on an earlier woodcut by Hans Grave, shows it as a
neatly laid out city that had not yet outgrown its

fourteenth-century walls (some districts would have
actually been a good deal more cramped). Like many
of the most effective plates in the *Civitates*, the map is
a compromise between the diagrammatic ground plan
and the pictorial view. Streets and buildings are
positioned more or less accurately, but the bird's-eye
view has enabled the artist to show the elevations of
the buildings, delineating prominent features such as
spires, towers, and bridges, and creating a map that is
visually attractive as well as useful.

LOCAL CHARACTERS
Foreground vignettes of city
dwellers in local costume are
an enlivening feature of the
Civitates. Here, they are a
soberly dressed middle-class
couple and a working woman
carrying produce into the city.

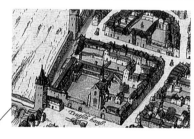

TEUTONIC KNIGHTS
This imposing complex, called the
Deutschordenhaus, was built by the Teutonic
Knights. From 1219 this crusading order was
settled on the south bank at Sachsenhausen,
which in 1318 became a suburb of Frankfurt.

PROSPEROUS CITY
The line of the twelfth-century walls
and moat is still visible, but the city
has greatly increased in size and
new and elaborate defenses guard
it—evidence of the city's prosperity
and expansion.

MIGHTY CHURCH
The 312-ft.-high (95 m) tower
of the cathedral of St.
Bartholomaüs is correctly
shown as dominating the city.
From 1562 the Holy Roman
emperors were crowned here.

EARLY TOWN PLANS

Skillfully made town views and plans were produced in the ancient world, only to disappear for centuries. Then commercial expansion, geographical discoveries, and the invention of printing laid the groundwork for a flourishing urban culture, vividly reflected in Braun and Hogenberg's masterly volumes.

The impulse to map the spatial relationship between human dwellings probably dates back to the earliest settlements of any considerable size. It can certainly be traced to the seventh millennium B.C., when a wall at Çatal Hüyük in Anatolia (present-day Turkey) was painted with a plan representing about eighty houses. From 2300 B.C. city plans were incised on clay tablets at Babylon, and at least one later example may have been drawn to scale. Roman surveyors were undoubtedly skilled at mapmaking, although only fragments remain of a great map of Rome, carved in stone, that recorded the entire layout of the city's streets.

Like other types of maps, the urban plan virtually disappeared during the Dark Ages, when many skills were forgotten and the role of towns diminished. Even when a flourishing medieval culture developed, between the eleventh and fourteenth centuries, most European maps indicated cities and towns with nothing more than single picture symbols. The main exceptions during this period were in Italy and Palestine. The Holy Land was open to pilgrims from 1099, when Crusaders captured Jerusalem, until its final loss in 1244. But although many maps of Jerusalem were made, most were circular and diagrammatic, without significant geographical content.

In Italy the tradition of urban culture was particularly strong. Apart from an exceptional tenth-century plan of Verona, the earliest known urban maps date from the twelfth century, beginning with Rome and Venice. Such Italian achievements grew out of a mapmaking tradition. By contrast, the remarkable Albertinischer plan, showing Vienna and Bratislava, appears to have been an isolated feat. Believed to be a mid-fifteenth-century copy of a 1422 original, it is the first European town map that is explicitly drawn to scale.

Above: Braun and Hogenberg's bird's-eye view of Leipzig in the late sixteenth century. In what is now Germany, towns had emerged as important centers of commerce; many had also achieved virtual independence as imperial free cities.

The fifteenth-century revolution in printing turned mapmaking into a potentially profitable occupation, especially as trade and travel increased. Nevertheless, most early printed books still offered views rather than plans of any featured towns. The lavishly woodcut-illustrated *Nuremberg Chronicle* (1493) contained the first printed view of what was labeled as an English town, though the woodcut was probably purely imaginary. The accompanying text suggests that the authors thought of England as a remote and half-fabulous place.

The pioneers

In the sixteenth century, many individual town maps and views were printed in the form of woodcuts or engravings. As early as 1500, Jacopo de'Barbari produced a pioneering work of great beauty and astonishing accuracy in his woodcut of Venice from a very high viewpoint. The most important precursor of Braun and Hogenberg, however, was Sebastian Münster, whose *Cosmographia* (1544) contained some sixty plans and views.

Braun and Hogenberg drew freely on earlier publications, but their own efforts were on an altogether grander scale. Their work—the first town atlas—comprised 530 town plans from four continents and represented extended research, correspondence, and travel, both by Braun and by the artist Joris Hoefnagel, whose contribution to Braun and Hogenberg has been stressed by recent scholarship.

Braun, the chief editor, was a canon of Cologne Cathedral, and the *Civitates* was printed in the city. Though its volumes contained examples of simple town views and basic ground plans, the most striking plates were the bird's-eye views mixing elevations and plans. These, picking out civic monuments and accompanied by examples of local costume and heraldic details, made the *Civitates* flattering to civic feelings and doubtless helped to sustain its popularity. Aspects of reality that do not appear in its pages are slum areas or hints of destructive sieges and sackings.

A long-lived masterwork

Braun and Hogenberg's work quickly found imitators, and new plans and views were made as city life expanded. Yet further editions of the *Civitates* continued to appear. In the 1650s, long after the deaths of its creators, a well-known Dutch map publisher, Jan Jansson, acquired the plates and published the work again. A final edition appeared around 1750, before time overtook Braun and Hogenberg's masterwork.

Above: The *Nuremberg Chronicle* of 1493 contained numerous woodcuts of towns, such as this one of Jerusalem. The view owes more to imagination and artistic convention than to reality. Religious buildings, both Christian and Muslim, are given great prominence.

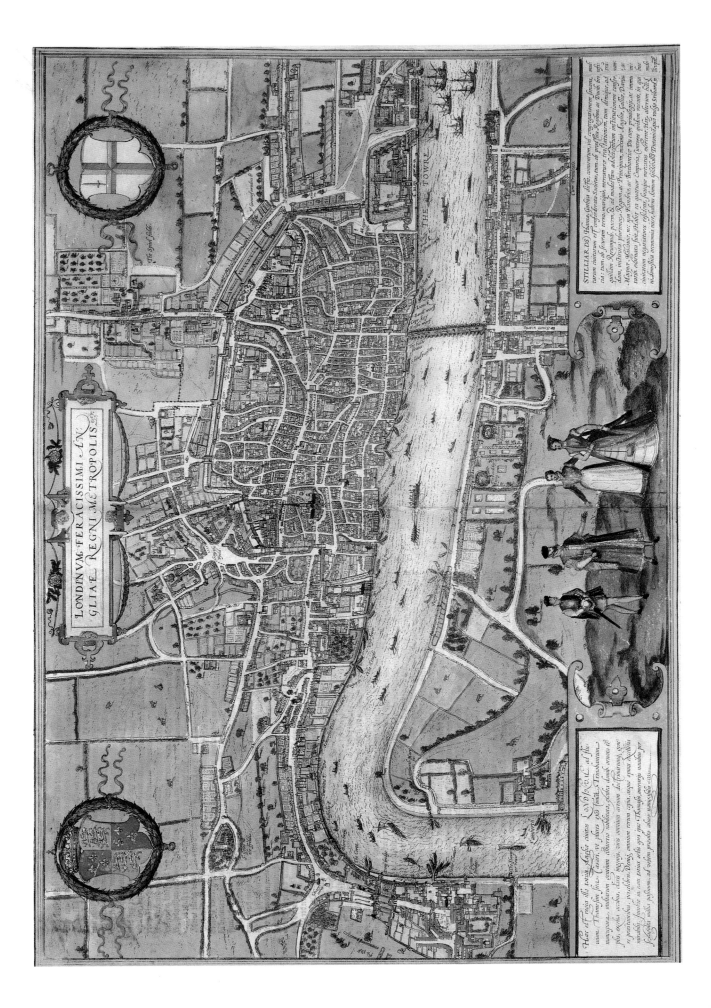

LONDON
BRAUN AND HOGENBERG

Described in its Latin heading as the "Capital of the Most Fruitful Kingdom of England," this plan of Tudor London was the first plate in Braun and Hogenberg's *Civitates*. Though surrounded by green fields, the city was already exceptional in size and spilling over into neighboring areas.

Issued in 1572, the plan was based on a large earlier map of the 1550s. It shows London as it was just before the accession of Queen Elizabeth I. London proper stands on the north side of the River Thames. It is distinct from the City of Westminster, home of the court and parliament, which lies to the west. The two cities are linked by the Strand, a thoroughfare lined with the grand houses of bishops and other notables. To the south of the river lies the suburb of Southwark, beyond the control of the city fathers. It offered dubious forms of entertainment, notably bull- and bear-baiting rings (marked) and brothels (unmarked), which would be joined in a few years by equally frowned-upon structures—theaters, including Shakespeare's Globe. The city plan catered to the mercantile interests of Braun and Hogenberg's mostly Germanic clientele with portraits of solid citizens in the foreground, details of the city such as tiny drawings of the water pump and stalls in busy commercial Cheapside, and informative corner panels.

ST. PAUL'S
London's cathedral appears as it was before the tall spire was destroyed in 1561. St. Paul's was a center of business, as well as worship, until the Royal Exchange was built in 1571.

CRUEL SPORTS
The bull- and bear-baiting houses of Southwark are shown, their arenas open to the sky. Within a very few years the same grounds would house theaters, in one of which Shakespeare's works were performed.

FAMOUS BRIDGE
London Bridge was the only bridge across the Thames until 1749. Its many arches and the houses that stood on it (making it like "a beautiful street," according to the bottom-left panel) are clearly visible.

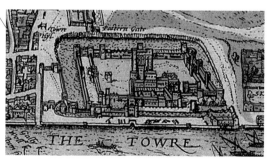

TOWER OF LONDON
The tower originated as a Norman fortress, intended to overawe Londoners. Later it was used to hold state prisoners, few of whom, once they had been rowed in through "Traitor's Gate," were ever at liberty again.

LONDON RISING

By 1572, when Braun and Hogenberg's map was published, London was already preeminent among British cities. The imperatives of geography, commerce, and government all conspired to give it an exceptional place in the country's life from Roman times to the postindustrial age.

Stone Age people left some of their flint axes on the site of London, but it was soon after the Roman conquest of southern Britain in A.D. 43 that a permanent settlement of any size grew up there, around the lowest point on the Thames at which the river could be forded. In 60 Londinium was burned to the ground by a revolt of the Iceni, led by Boudicca, but it was rapidly rebuilt and became the largest city, the greatest port, and the hub of the road network in Roman Britain. The Thames was bridged at an early

date, the city was walled about 200, and in 286 a mint was established. Until the Romans departed, London was a sophisticated urban center with amenities including central heating and an amphitheater.

Eclipsed during the post-Roman period, London recovered some of its standing in the less urbanized society of the Anglo-Saxons. After the Norman Conquest in 1066, King William I regarded the security of the city as important enough to build the White Tower (now part of the Tower of London) to protect and overawe Londoners. The earliest long description of the city was written in the 1180s by William FitzStephen, who described in admiring terms its bustling trades and the way Londoners relaxed by skating on the ice at Moorfields. In the Middle Ages, London established its municipal independence and acquired a lord mayor and flourishing mercantile and craft companies and guilds.

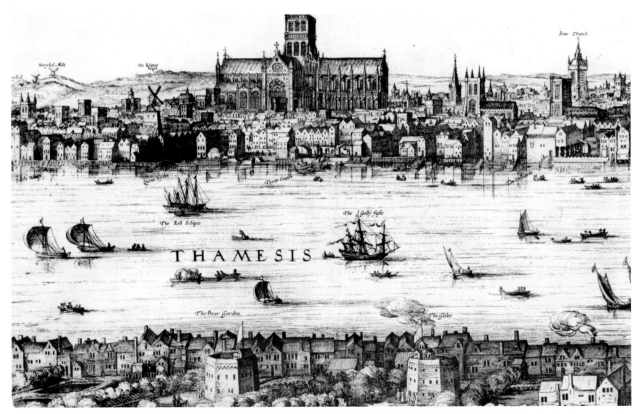

Above: Before the fire: This image of London, drawn in 1616, shows St. Paul's Cathedral dominating the skyline. In the foreground stand the Globe Theater and the Bear Garden in Southwark, the traditional location of London's more risqué leisure pursuits.

PROSPECT OF THE CITTY OF LONDON, AS IT APPEARED, IN THE TIME OF ITS FLAMES.

Southwarke

Above: Flames and smoke engulf the north bank of the Thames in this contemporary engraving of the Great Fire of London, which broke out on September 2, 1666. Just left of the center of the image is Old St. Paul's Cathedral, which was destroyed in the fire.

The expanding city

In the sixteenth century London began to grow rapidly, and by the century's end Braun and Hogenberg's plan, reflecting the situation in the 1550s, was definitely behind the times. In particular, urban development was rapidly filling up the area between the City of London and the previously separate City of Westminster, so that the court at Whitehall and the Houses of Parliament effectively became part of a larger London. However, the condition and layout of the streets was such that the Thames remained the main metropolitan highway, and the watermen who plied up and down the river were an independent and characterful breed, not unlike a later London type, the taxi driver. In addition to the Braun and Hogenberg plan, the Tudor city was recorded in admirable detail in *A Survey of London* (1598–1603) by John Stow, a craftsman-antiquary of the same stamp as John Speed.

The linking of the City and Westminster had political repercussions in the seventeenth century, when Londoners played an important part in supporting Parliament against Charles I. The most dramatic episode in the city's history, however, was the Great Fire of London in 1666, which followed hard on the horrors of the plague the year before. Most of the city was destroyed, including more than 13,000 dwellings, St. Paul's Cathedral, and eighty-seven churches. The catastrophe gave Sir Christopher Wren the opportunity to build a new St. Paul's and fifty-one churches in the then-modern classical-Baroque style. However, Wren was never able to carry out his plans for an entirely new, planned London of broad, radiating avenues and pleasant vistas.

London was rebuilt in a modified form of the earlier irregular layout. The rational planning envisaged by Wren was implemented elsewhere, at the "West End," during the Georgian and Regency periods when elegant terraces and villas were built over a wide area from St. James's to Regent's Park and beyond. From the 1800s Britain's commercial and maritime preeminence was reflected in great dock construction schemes, undertaken in the Isle of Dogs and other parts of what was to become the East End. And after 1749 London at last had more than a single bridge spanning the Thames.

This was the prelude to an even greater nineteenth-century expansion, an extraordinary record of slum and suburban building, over- and underground railways, epidemics, and public works, which effected the transformation of London into a megalopolis of some six million people.

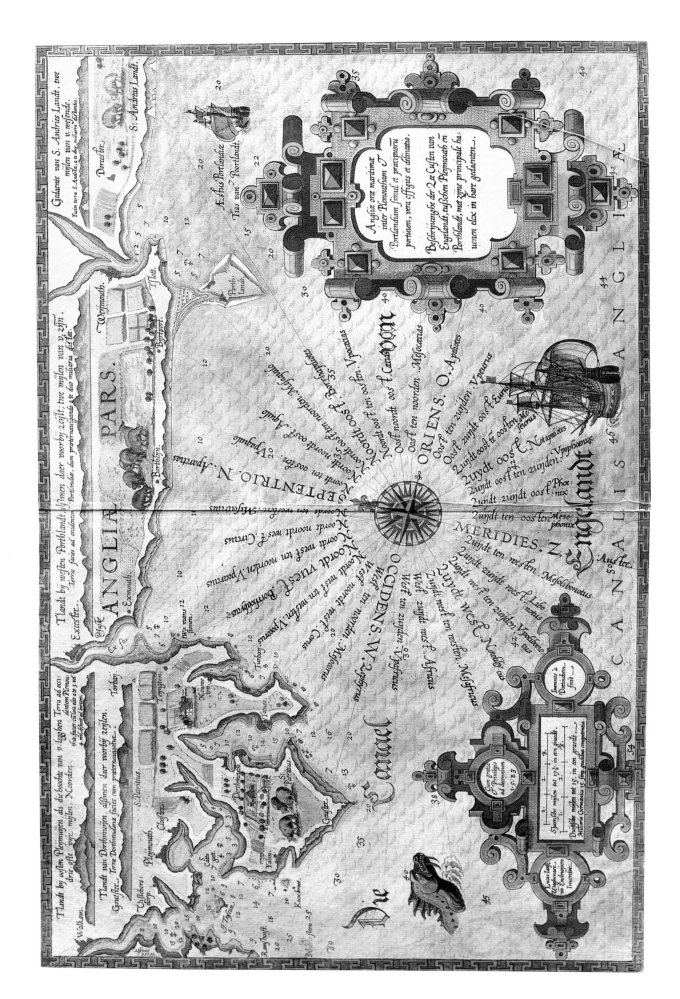

THE ENGLISH CHANNEL
LUCAS WAGHENAER

From the seventeenth century English mariners described sea charts as "waggoners"—a tongue-twisted tribute to the Dutch cartographer Lucas Waghenaer, who published the first printed atlas of charts. Splendidly decorated, the atlas was also groundbreaking in its use of innovative symbols and wealth of hydrographic information.

After having worked as a pilot, Waghenaer became an excise officer at his birthplace, Enkuizen on the Zuider Zee, in 1579, but three years later he was accused of accepting bribes and dismissed. Guilty or not, he was exonerated after the great success of his *Spieghel der Zeevaert* ("Mirror of Navigation"), published in 1584–1585. Until the publication of the *Spieghel*, northern mariners had relied on written sailing directions and word of mouth. The forty-four charts of Waghenaer's work provided a record of the coastal waters of Europe from Gibraltar to the Baltic, accessible to all through the medium of print. As well as employing familiar portolan symbols for rocks and shallows, Waghenaer devised a range of new symbols, many of them still in use today, showing depth soundings, shipping channels, and tides, and included views of coastlines from a few miles out at sea, a useful aid to navigation. Waghenaer produced a new chart book, *Thresoor der Zeevart* ("Treasury of Navigation"), in 1592.

SEA MONSTER
The inclusion of a sea monster on an up-to-date chart of familiar waters might seem like an anachronism, but such creatures long remained a decorative feature of maps and charts.

RUSTIC SCENES
Science and scenery: below the long, utilitarian profile of the English coast between the River Exe and Weymouth, the artist has placed vignettes of peaceful rural life.

VITAL PORT
A very large number of depth soundings are recorded in the approaches to Plymouth, an English port of vital strategic as well as commercial significance.

SCALE BARS
A splendid strapwork cartouche holds the scale bars, which in this first (Dutch) edition are expressed in Dutch and Spanish miles.

WIND ROSE
A large and elegant wind rose dominates the center of the map, providing orientation and filling a space of no great relevance to a chart of coastal waters.

INTO THE AGE OF FIGHTING SAIL

The English translation of Waghenaer's book, entitled *The Mariner's Mirror*, was published in 1588. Ironically, although the book met a definite need, the year was one in which English seafarers proved the quality of their skill and knowledge by defeating Spain's "Invincible Armada" and inaugurating a new epoch of naval warfare.

Above: Portuguese mariners use a range of instruments to steer their warship in this sixteenth-century engraving. The "Spanish" Armada included a large contingent of Portuguese vessels.

The Armada was an invasion fleet of 130 ships sent by Philip II of Spain to conquer heretical England. If it succeeded, Spain and the Catholic cause it championed would have good prospects of dominating Europe. Believing that he was carrying out God's will, Philip coined the term "Invincible Armada" to describe his fleet.

His commanders were less confident. They recognized that the English galleons, even though they were smaller than their own vessels, were more maneuverable and were equipped with superior long-range firepower. The English were also experienced seafarers, both in home and transatlantic waters, and the Spaniards had already suffered much at the hands of the fiery piratical Sir Francis Drake. Nevertheless, Philip's enterprise did have the benefit of Portuguese ships and seamanship, for his "Spanish Armada" was, in fact, a combined fleet.

An unpredictable outcome

Although the odds favored the English, the hazards of the weather in northern seas made the outcome of the naval campaign impossible to predict—all the more so because there had been nothing like it before. Since ancient times great sea battles had been fought in the tideless Mediterranean by fleets of galleys, but they were propelled mainly by oars, and combat involved an adaptation of land tactics: The attacking ship drew alongside its prey and used ropes and hooks to bind the two vessels together (grappling). Then the enemy was boarded and the crews fought hand to hand.

Galleons, on the other hand, were adapted to ocean voyages. Their decks lined with cannon, they fought at a distance, maneuvering for advantage and

delivering and receiving devastating broadsides. Engagements had so far been quite small scale: The unpredictable campaign of 1588 marked the beginning of the age of fighting sail.

From the outset the winds were capricious, holding back the Armada and then driving it forward just when the waiting English fleet was taking on more provisions in Plymouth harbor. The Spaniards might have seized the opportunity to trap and destroy their enemies, but they obediently followed the inflexible plan laid down by King Philip. In no circumstance, according to Philip's instructions, was the Spanish fleet to initiate action. Its task was simply to pick up the Spanish army in the Netherlands and ferry it across the English Channel to invade England.

The Armada sailed on, shadowed by the English. During the early clashes, the English tactics were to use their superior seamanship and maneuverability to stay out of range while pounding the Spaniards from a reasonably safe distance. The engagements were fierce, using up unprecedented quantities of powder and shot. Although the English were always in the ascendant, however, long-distance gunnery proved far less effective than both sides had imagined, and Spanish losses were relatively light. The Armada arrived at its rendezvous, not far from Calais, France, a little battered but in good order.

King Philip's plan for a liaison with the Spanish army was deeply flawed, but the English were not to know that. Reinforced, they had—contrary to legend—more ships than the Spaniards and were close enough to their home base to replenish their stocks of ammunition. On the night of August 7–8, to break the enemy's close order, they sent in eight fire-ships—vessels that had been packed with gunpowder and iron and deliberately set alight. The tactic was not unknown, but the English had sacrificed larger vessels than usual. Failing to fend them off, the Spaniards cut the cables that anchored them in position and scattered into the night.

At seven in the morning, the English attacked and the final encounter began. The Battle of Gravelines consisted of a series of engagements in which English ships swarmed around isolated Spanish galleons, pouring in broadsides, while other Spanish vessels tried to rescue their comrades and regroup. When a squall ended the fighting late in the afternoon, many of the Spanish ships were shattered, although it was evening before one sank and two more were beached.

Armada destroyed

The English victory remained less complete than might have been expected. However, strong winds then drove the Armada north, forcing it to sail home around the British Isles. The wild Atlantic and the rocky west coast of Ireland turned defeat into catastrophe, sinking and wrecking vessel after vessel. In the end, about sixty of King Philip's ships were lost after the first great sea battle of the age of sail.

Above: The Royal Navy dockyard at Portsmouth, on Britain's south coast. Portsmouth has been a naval station since the thirteenth century and a major center for the Royal Navy for the last 500 years.

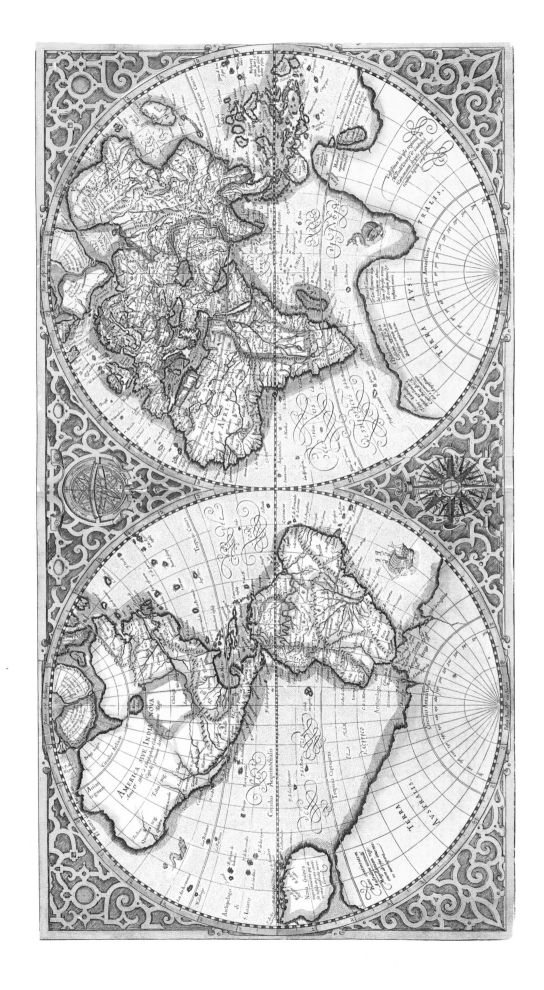

WORLD MAP
RUMOLD MERCATOR

Mercator is without doubt the most famous name in cartography.
Gerardus Mercator was a master mapmaker, coined the term "atlas," and invented the
projection that has since been used for many thousands of terrestrial maps. Rumold
Mercator completed his father's atlas and produced this world map, based on Gerardus's
extremely rare work of 1569.

By his death in 1594, Gerardus Mercator had issued two parts of his great atlas. Rumold was closely involved in his father's project. He went on to complete the third part, and published the first complete edition of the atlas in 1595. His own world map is dated 1587, which suggests that it was sold separately before being included in the final work. Though similar to Gerardus's map (including the rendering of the north polar regions as a group of islands), it is not a copy. And unlike the 1569 map, Rumold's work is in the double hemisphere form pioneered by Italian cartographers; Girolamo Ruscelli's 1561 edition of Ptolemy was the first collection to include such a map. The double hemisphere left large spaces that Rumold filled with elegant strapwork, an abstract equivalent to the lavish decorative borders that appeared in the 1590s and remained in vogue for more than 200 years.

MYTHICAL KINGDOM
Anian, a kingdom mentioned by Marco Polo, became a potent myth. Its equally imaginary strait offered the longed-for northwest passage to the East. Hence America is shown, correctly, as separate from Asia.

ARCTIC ISLANDS
Proving that the age of imaginary mapmaking was far from over, Rumold Mercator followed his father in filling the Arctic with islands, along with descriptions of their strange properties and inhabitants.

SOUTHERN CONTINENT
Tierra del Fuego appears as part of the vast southern continent, Terra Australis; a passage around it was not discovered until 1616. Though New Guinea is shown as an island, the inscription leaves the matter in doubt.

JAPAN
On this map the Japanese archipelago has finally been located in roughly the right place, but it takes the form of a single large island. There is no sign of Korea.

MERCATOR'S PROJECTION

Rumold Mercator was the capable son of an even more distinguished maker of maps, Gerardus Mercator. He devised the most popular of all map projections, predominant for centuries and still used for nautical charts.

Gerardus Mercator was born Gerhard Kremer in 1512. Like many sixteenth-century scholars, he Latinized his name, incidentally giving himself added distinction by turning Kremer (shopkeeper) into Mercator (merchant). The son of poor German parents who had not long before moved to Flanders, he managed to study at Louvain University and acquired a thorough knowledge of geography, astronomy, and surveying before setting up as a cartographer and instrument-maker.

In 1552 Mercator moved from Louvain, probably because his Protestant convictions had caused him difficulties, and settled in the German city of Duisburg. During his long career, he composed religious works and a handwriting manual, as well as maps, including a famous double cordiform (heart-shaped) world map (1538), a map of the British Isles (1564), and the first two parts of the atlas completed by Rumold in 1595, a year after Gerardus's death. Mercator's choice of *Atlas* for his title alluded to a mythical being, one of the Titans who rebelled against the gods. Originally Atlas was said to have been condemned to support the sky on his shoulders, but in later myth his burden became the world. It was this appropriate image that made "atlas" an attractive term with which to describe a collection of maps—although Mercator's actual intention in using the word had been more learnedly obscure.

The Mercator projection

In the eyes of posterity, the greatest of all Mercator's achievements was his 1569 sheet map of the world, drawn to his famous projection. He was not the inventor of the projection, but he was the first maker to print a world map that was based on it and was consciously designed to assist navigation at sea.

Like all mapmakers, Mercator was faced with the impossibility of transferring features undistorted from a curved globe on to a flat surface. Each method used—projection—achieved accuracy in one respect while creating distortions in others. As pictured by Mercator's projection, landmasses became grossly distorted toward the poles, while the poles themselves could not be represented at all.

This resulted from the way in which Mercator had devised the map, covering it with straight parallel lines of longitude and latitude that formed a rectangular grid. The great virtue of this arrangement was that the correct angular relation between places was preserved. What this meant in practice was that a straight line drawn from A to B on a Mercator projection map could be followed across the sea without any need for constant adjustments of direction, which had previously been unavoidable. In more technical language, the navigator who used a Mercator chart was able to maintain a constant bearing. Owing to the curvature of the Earth, lines drawn on Mercator's map were not the very shortest distances between two points (true direction), but the errors were relatively

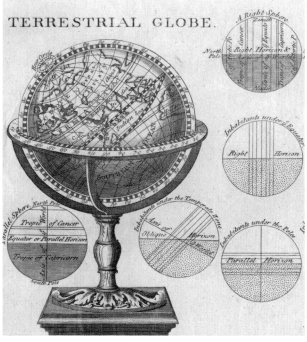

Above: *The earliest surviving globe dates from 1492—this example was engraved around a century later.*

small and were far outweighed by the convenience of being able to plot an unchanging course.

The immediate results of Mercator's innovation were almost nonexistent. Conservative-minded seamen saw no reason to give up their familiar manuscript charts and sailing directions, especially since Mercator's projection did not provide accurate information about sailing distances, the traditional way in which sailors planned their voyages.

Part of the responsibility for this lay with Mercator himself, who had not explained his projection clearly.

Above: Published in 1595, Gerardus Mercator's *Septentrionalium Terrarum Descriptio* was the first map devoted to the Arctic. He shows the North Pole as a group of rocky islands in the Arctic Ocean with four channels separating them.

This was remedied in 1599, when the English mathematician Thomas Wright published a version of Mercator's map that was more easily understood. Even so, it was not until the eighteenth century that Mercator's projection was universally recognized for what it was—indispensable to the navigator.

Septentrio.

Arcipelago di S. Lazaro.

Baixos de S. Bartholome
I. de S. Petro.

OCCI-

DENTALIS.

La de Sagrou chada
Corral de Praceles
I. de Praceles
Barbudos
I. de Dcu Alonco

De lar dos Teç
I. de Puercos
I. de los Nadadores

Miracomo Vel

Iſlas de Salamon.

Nova Guinea.

Sic a nautis dicta, quod
littora illa, conditio̅q̅ terræ,
Guineæ in Africa multum
ſimilia ſint. Continens̅ ne
ad terra̅ Auſtrale̅, an In-
ſula ſit, incognitu̅ eſt.

Nombre de Iheſus
Baixas de Candelare
Malarts
De la Aguada
S. Xpoual
S. Thiago

Lamba
I. de los Tabarones

Velle de Loua
Guerre
I. Dupoa
Delcas

S. Nicolai

Tropic. Capricorni.

Oriens.

Meridies.

NOVÆ GVINEÆ
Forma, & Situs.

NEW GUINEA
GERARD AND CORNELIS DE JODE

The late sixteenth century was one of the great ages of cartography, centered on the Low
Countries. Though overshadowed by Ortelius and Mercator, Gerard de Jode and
his son, Cornelis, produced excellent atlases and individual maps, including
unusual details of newly discovered lands.

Gerard de Jode began publishing maps at
Antwerp in Belgium in 1555. In the 1570s de
Jode worked on his own atlas, *Speculum Orbis
Terrarum* ("Mirror of the World"), issued in 1578 with
sixty-five maps. His son, Cornelis, collaborated with
Gerard in preparing an expanded version of the atlas
with 112 maps. It appeared in 1593, two years after
Gerard's death. The unusually detailed accounts of
North America suggest a lively interest in contemporary
discoveries, which also appears in this early (and
highly speculative) map of New Guinea. An
inscription notes that the island took its name from a
fancied resemblance between the north coast and the
Guinea coast in West Africa. It also states that it was
not known whether New Guinea *was* an island or part
of Terra Australis, the southern continent. The region
shown below New Guinea is not an intuition of
Australia's existence, but part of the southern
continent. De Jode's drawings of its outlines and
inhabitants were, of course, purely imaginary.

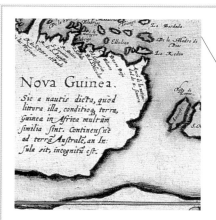

NEW GUINEA
New Guinea is correctly shown as an
island, but is not accurately rendered
in most other respects. The inscription
expresses de Jode's uncertainty about
its geographical status.

HAPPY COUPLE
The embracing mermaid and merman
are the most agreeable of the imaginary
creatures with which the waters have
been half-seriously peopled—a
convention that would long outlive
terrestrial monsters on maps.

DANGEROUS CREATURES
De Jode's imagined
southern continent contains
a dragonlike creature with
wings, worthy of a medieval
cartographer. Though also
confronted with a lurking
lion and snake, the
"Australian" bowman
appears to be undaunted.

FOUND AND LOST
The Solomon Islands are
correctly placed close to
New Guinea. Mendaña
discovered them in 1568,
but in 1595 (two years
after this map was
published) he was utterly
unsuccessful in finding
them again.

SEEKING THE SOUTH

For centuries Europeans were fascinated by the imagined wonders of a hypothetical southern continent, and after 1600 they made a number of attempts to locate it. Ironically, when they did find a continent in southern latitudes—Australia—they were sadly disappointed.

The ancient Greeks believed in the existence of a southern continent—Terra Australis, or the Antipodes. Aware that the Earth was a sphere, they argued that, for the sake of stability, there must be a southern landmass to balance that of Eurasia. Ptolemy described this southern continent as joined to Africa and extending to the east, making the Indian Ocean an inland sea. Terra Australis came to be thought of as a place of marvels, including peaceful inhabitants, unicorns, and fabulous wealth—all impossible to verify, since a blood-boiling torrid zone was believed to prevent travel from north to south.

The Church rather disapproved of the notion of unreachable, unbaptized races, but the idea of a southern continent survived the Middle Ages and revived wonderfully once Portuguese explorers had passed unharmed through the torrid zone.

The search was slow to begin. Although the Portuguese were sailing close to the north coast of New Guinea by 1540, they were too busy exploiting their East Indian empire to go farther. When Spanish ships entered the Pacific from the east, they met winds that blew them north, away from Terra Australis.

Nevertheless there were some near misses. In 1568 Alvaro de Mendaña discovered the Solomon Islands. In 1605 Pedro Fernandez de Quirós landed on what he believed to be the southern continent but was in fact one of the New Hebrides group. He named it La Australia del Espiritu Santo and founded what was meant to be a New Jerusalem, before the lack of gold and a mutinous crew forced him to move on. Closest of all was Luis Vaez Torres, who navigated the strait named after him, unaware that the only habitable southern continent—Australia—lay just to the south.

Below: The southern continent, Terra Australis, was rumored to be the home of all manner of fantastical creatures, such as this maned unicorn drawn around 1500.

The navigators who finally found the southern continent were hardly more conscious of the fact. In 1606, the year of Torres's voyage, Willem Jansz was sent on a gold-prospecting mission by the Dutch authorities at Batavia (Jakarta), who had replaced the Portuguese as the dominant European power in the East Indies. While dutifully heading for what he believed to be the coast of New Guinea, Jansz reached the Gulf of Carpentaria and made the first-known European landing on Australian soil. Finding no gold and uncongenial natives, he returned to report without enthusiasm on his discoveries.

Australian landfalls

Nevertheless, there were a number of contacts in the years that followed, mainly because the Dutch discovered a faster route to Batavia, sailing east from the Cape of Good Hope and turning north only late in the voyage. Vessels following the route that mistimed their turn made landfall—or were wrecked—on the shores of Australia.

This happened often enough for the Dutch to become familiar with the west and southwest coasts of "New Holland." The next major advance was made in

1642–1643, when an unusually forward-looking governor-general at Batavia, Anthony Van Diemen, sent Abel Tasman to explore as far south as he could. Leaving Mauritius, Tasman sailed south of the Australian coast and landed on what he named Van Diemen's Land (the island of Tasmania). When the Tasmanian coast began to turn away to the north, Tasman decided to continue sailing east. He reached and rounded New Zealand, and returned to Batavia via Tonga and Fiji.

During a second voyage in 1644, Tasman failed to prove that New Guinea was an island by finding the Torres Strait. But he did thoroughly explore the northern Australian coast from the Gulf of Carpentaria to North West Cape. It was now clear that Australia was a large island, although its east coast was not yet known and some believed that it stretched as far as New Zealand. It was also clear to the Dutch that the climate was arid and the natives knew nothing of gold or other riches. After Van Diemen's death in 1645, they lost interest. Despite expeditions by William Dampier and others, puzzles concerning the South Land would persist until the late eighteenth century, when Captain Cook solved them all.

Above: The central offices of the Dutch East India Company in Bengal, India, pictured in about 1665. The Company's dominant role in the Spice Islands of Southeast Asia meant that Dutch navigators made contacts, by accident as well as design, with Australia.

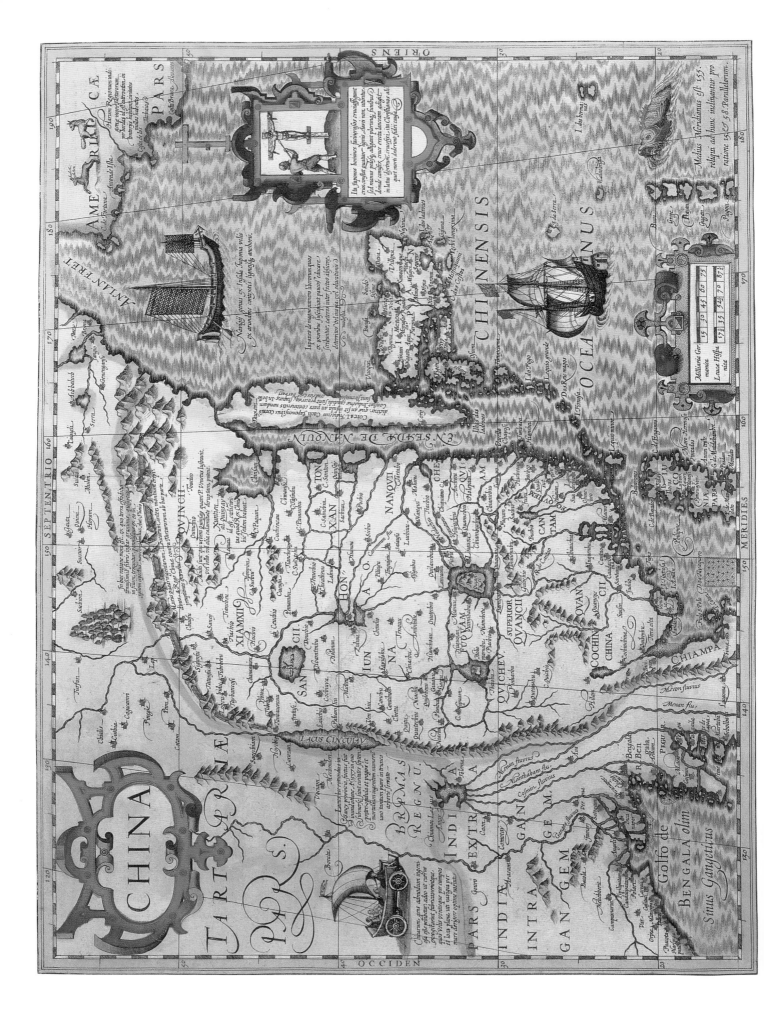

CHINA
JODOCUS HONDIUS

At the beginning of the seventeenth century, European knowledge of China and the Far East had made considerable advances since Marco Polo's day, but it remained very patchy. Hondius's map of 1606, packed with information and exotic images, is a reliable compendium of contemporary facts and fantasies about China and its neighbors.

Hondius was born Jost de Hondt at Ghent in Flanders. About 1583 he settled in London, where he worked as a cartographer and engraver, making the plates for *The Mariner's Mirror* (1588), the English version of Waghenaer's sea charts. Two years later he produced a striking double-hemisphere map illustrating Francis Drake's voyage round the world in 1577–1580. About 1593 Hondius moved to Amsterdam and set up his own business as a publisher of maps and globes. His links with England remained strong, and he engraved a set of twelve county maps for William Smith in 1602–1603, and all the plates for John Speed's *Theatre of the Empire of Great Britain* (1611). In 1604 he bought the engraved plates of Mercator's atlas. Two years later he reissued it, in greatly expanded form, in the first of many new editions. On the map illustrated here, America is too close, and China, Korea, and Japan are squeezed together, but the information was the best available and the engraving is of very high quality.

KOREAN ISLAND
Like his contemporaries, Hondius shows Korea as an island, though he expresses his uncertainty in the inscription. In 1631 his son Henricus depicted it, correctly, as a peninsula.

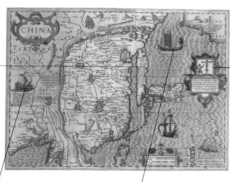

CRUCIFIXION
According to the inscription, the Japanese tortured and executed malefactors in a fashion resembling crucifixions in Christian tradition. Persecution of Christians had occurred in Japan and later became severe.

LAND SHIP
Though Chinese civilization had lost some of its dynamism, Europeans were still impressed by rumors about its more ingenious inventions, including this horseless carriage driven by the wind.

JAPANESE JUNK
This is a reasonably convincing picture of the type of junk sailed by the Japanese; the inscription notes that it is equipped with "sails woven from reeds and with wooden anchors."

WESTERN STRANGERS

Hondius's map of China showed that
Europeans were beginning to learn
something about the East after several
centuries in which little had been added
to the famous account by Marco Polo.
The new information came almost
exclusively from missionaries who strove
to convert the Chinese but were also
capable of accepting a great civilization
on its own terms.

Western knowledge of East Asia had stagnated
after the collapse of the "Mongol peace,"
during which Europeans had been able to
cross Asia in relative safety. As a result, Marco Polo's
early fourteenth-century *Description of the World* and
the writings of friars who had tried to convert the
Mongols to Christianity remained the authorities on
Tartars (Mongols), and on China, for almost three
centuries. On Abraham Ortelius's 1570 map of
Tartary (see page 104), the Great Khan continued to
reign over an empire that had long since broken up
into rival khanates, and one of the frontiers of the
Tartar realms was said to be the river Caromoran
(the Yangtze), even though the Chinese had expelled
the Mongols as long before as 1368.

Portuguese trade with the East
Contact between the West and China was renewed
after the Portuguese succeeded in rounding the Cape
of Good Hope and finding their way to India and the
East by sea. By 1514 they had reached China, and after
some vicissitudes they acquired the port of Macao and
were allowed to trade from there. Relations were so
carefully regulated by the Chinese, however, that
Portuguese merchants and seamen learned little about
the land they did business with.

By contrast, intrepid Portuguese Jesuits managed to
travel more widely and observe more carefully. The
Jesuits combined missionary fervor with an interest in
science and a willingness to adapt to other cultures.
An early byproduct of their Chinese contacts was Luis
Jorge de Barbuda's *Description of China*, which was
accompanied by a map. This was used by Ortelius in
1584 and became the basis of subsequent efforts,

including the Chinese part of Hondius's work. On
Hondius's map China is shown as an independent
empire, as it had been since the Ming dynasty received
the Mandate of Heaven in 1368. There is also some
hint of its stature in the depiction of the Great Wall
and descriptions of Chinese inventions and products,

Above: This seventeenth-century engraving commemorates the
first convert of Jesuit missionary Matteo Ricci, on the left, whose
scientifc knowledge made him influential at the imperial court.

which Europeans were to appreciate for themselves as they became acquainted with such novel delights as tea, silk, and porcelain.

Matteo Ricci makes his name

Barbuda's book and map were important in the West, but it was Matteo Ricci, an Italian, who acquired a degree of influence for the Jesuits in China. After arriving at Goa in India in 1580, he studied and mastered the Chinese language and the Confucian classics with astonishing speed. From 1583 he worked in China itself, and from 1602 he was permitted to reside in Beijing. His knowledge of astronomy enabled him to correct the Chinese calendar, and as a result he was appointed a court official and was able to find similar employment on the Board of Astronomers for other learned Jesuits. He also produced a world map for the emperor, reviving a Chinese cartographic tradition that was far older than Europe's.

Ricci was succeeded by Adam Schall, whose correct prediction of an eclipse in 1624 led to an equally prestigious official appointment. Jesuit influence survived a change of dynasty when the Manchus captured Beijing, and it reached its height under the Emperor Kangxi (1661–1722). Though delighted by a Western novelty—a striking clock—Kangxi took the Jesuits seriously enough to employ them as advisers when he negotiated a treaty with Russia. He also encouraged Ferdinand Verbiest to complete a double-hemisphere map of the world, based on both European and Chinese sources, which was printed at Beijing in 1674.

The Chinese and Christianity

By 1692, when Kangxi issued an official edict of religious toleration, the first Chinese bishop had been consecrated and there were perhaps 200,000 Chinese Christians. However, the Jesuits owed their success to a controversial policy of "accommodations," in which the Confucian veneration of ancestors and its accompanying rites were regarded as social rather than religious practices, compatible with Christianity. In 1704, when the pope condemned the policy, an outraged Kangxi withdrew the edict and expelled many priests. Some Jesuits stayed on, but their

Above: The Great Wall of China, prominent on Hondius's map, was originally built in the third century B.C. to protect the empire from invaders from the north.

influence was greatly diminished. Jesuit writings and maps expanded European knowledge of China, but the country itself remained almost entirely closed to foreigners. Its opening up, by force, in the mid-nineteenth century began a long and unhappy chapter in Sino-European relations.

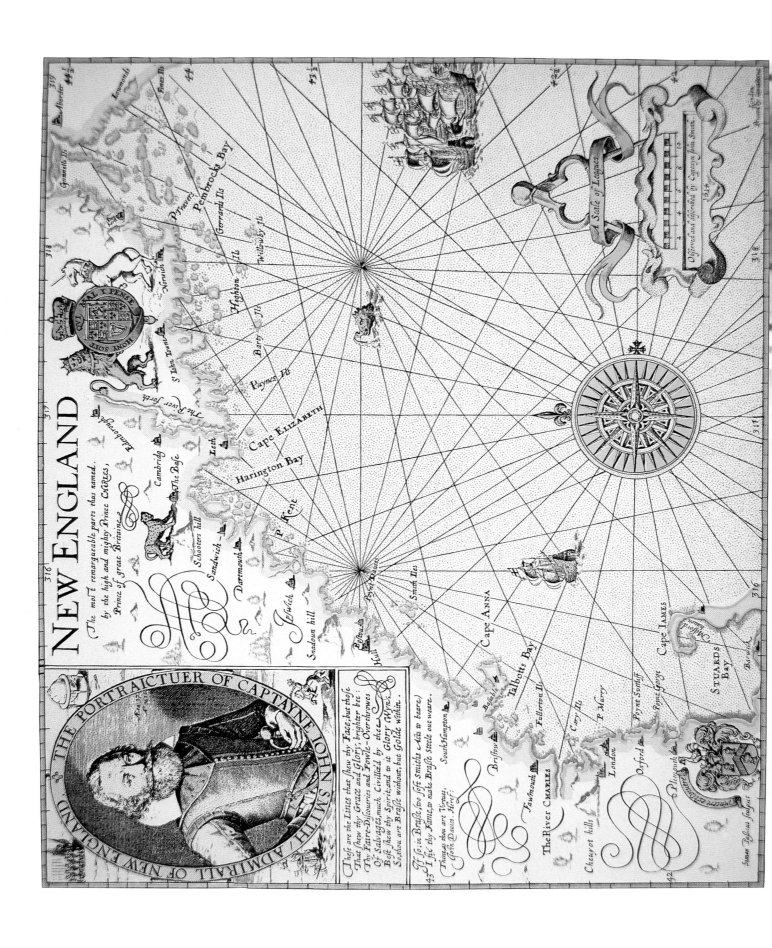

NEW ENGLAND
CAPTAIN JOHN SMITH

Smith was one of the heroes of early English colonization, playing a vital role in the first permanent settlement in the New World. He was also an explorer and gifted mapmaker, publishing accurate, decorative maps of Virginia and New England that were widely copied. He published the map illustrated here in 1616.

Smith led an adventurous life, fighting for Austria against the Turks, becoming a captive and slave, and seeing Constantinople and the Caucasus before making his way back to England. He joined the Virginia Company colonists bound for America, and in 1607 the colonists founded their settlement at Jamestown, Virginia. Smith undertook the first of two extensive explorations of Chesapeake Bay and its tributaries. He was captured by Native Americans in December 1607, but the intervention of the chief's daughter, Pocahontas, saved his life. In 1608 Smith became governor and instituted a strict regime that pulled the settlers through, famously insisting that "if any would not work, neither should he eat." In 1609 new dissensions and an injury caused Smith to leave for England, where in 1612 he published a general history and map of Virginia. After leading an expedition to New England, he published a *Description* of the region (1616) that included this map—aids that were used four years later by the Pilgrims.

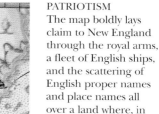

PATRIOTISM
The map boldly lays claim to New England through the royal arms, a fleet of English ships, and the scattering of English proper names and place names all over a land where, in fact, only Native Americans were living.

PORTRAIT
A self-assured figure, hand on hip and surrounded by self-flattering inscriptions and images (including his coat of arms, below, with three Turks' heads), Smith is clearly bent on making his mark as "the Admiral of New England."

CAPE
Cape James, named by Smith to flatter his sovereign, is now Cape Cod. Plymouth was a more successful naming, taken up by the Pilgrims. The arms are Smith's own, supposedly granted after he slew three Turks in single combat.

FLEET
An English fleet, the vessels flying flags bearing the cross of St. George: a symbol of English ambitions, but also a sight that Smith and other hard-pressed colonists must often have longed to see.

ADVENTURERS AND PILGRIMS

The first permanent English settlements in North America offered a contrast in human types and environments. Virginian adventurers and New England Pilgrims would be the earliest of the waves of immigrants who sailed west in search of a prosperity and freedom that had eluded them in the Old World.

After the failures of Queen Elizabeth I's reign, new colonizing ventures were launched in 1606 under King James I. Two royally chartered but merchant-funded companies were formed, a London, or Virginia, Company and a Plymouth Company that was licensed to establish settlements farther up the coast, above latitude 38 degrees north.

The Plymouth Company's early ventures were unsuccessful, and Jamestown in Virginia became the first permanent English colony, despite several perilous decades and a horrific death toll. This has often been attributed to the unsuitability of the colonists, who were gentlemen-adventurers and artisans rather than farmers. If their get-rich-quick attitude was reprehensible, however, it was encouraged by the company, whose instructions emphasized searching for a river passage to the East (still believed to be quite near) and prospecting for gold. And

whatever their shortcomings, the colonists also had to cope with an unprecedented drought that undermined their health and complicated their relations with their similarly afflicted Native American neighbors, the Powhatan Confederacy, on whom they depended for much of their food.

Under John Smith's leadership the colony weathered the first two years, but after his departure in September 1609 the winter that followed was so savage that it became known as "the starving time." Only sixty of the 500 colonists survived and, despite reinforcements, matters became so desperate that in June 1610 Jamestown was abandoned. It was saved by the coincidental arrival on the following day of a new governor who ordered the colonists to return.

Founding the tobacco industry

In 1614 Pocahontas, daughter of the Native American chief Powhatan, married John Rolfe, an alliance that gave the colony a few years' respite from Native American wars. Rolfe was an even more significant figure than Smith, since he was the first Englishman to grow tobacco in Virginia and export it to England. He laid the foundations of the industry that would make the colony prosperous and create its plantation character. For a time epidemics and wars with the Native Americans continued to threaten Jamestown, but by mid-century it was fully established and Powhatan's people had become the first victims of the

Right: Settlers in Jamestown, the first permanent English settlement in America, trade with local Native Americans. Initially good relations between the two groups deteriorated under increasing pressure from harsh weather and failed harvests.

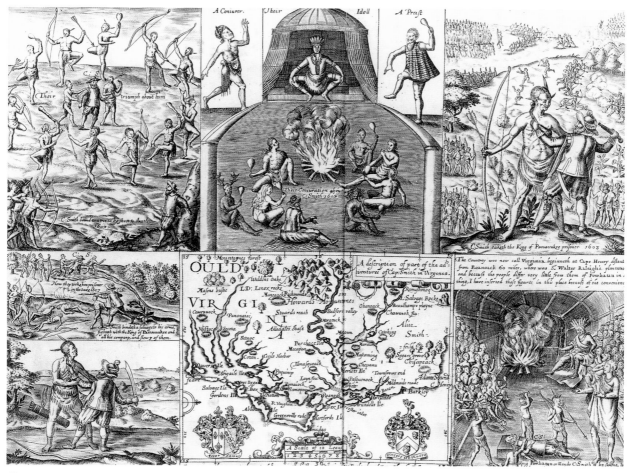

Above: This engraving from John Smith's 1642 *Generall Historie of Virginia* shows his adventures in the colony, including his capture and near execution by the chief Powhatan and the intercession of Pocahontas that saved him (bottom right).

European–American clash of cultures, consigned to a reservation. By contrast, the colonists, though subject to the Crown, managed to maintain their own assembly and run their own affairs.

The next colonizing venture was undertaken by the 104 men and women whom posterity has called the Pilgrims. All were deeply religious, and a large minority had broken with the Church of England and sought refuge in the more tolerant Netherlands. In 1620, having arranged for their passage with the London Company, they left England in the *Mayflower*. Instead of delivering them to company-controlled Virginia, however, the ship landed at Cape Cod. One effect of this accident was that the Pilgrims were free to govern themselves and drew up the celebrated "Mayflower Compact." Though always a small community, their priority of time and place—as well as their celebration of the first Thanksgiving—made the Pilgrims the epitome of the earnest, industrious

Puritan spirit that had such an influence on American life. Religious conflicts in England soon increased the Puritan presence. Massachusetts Bay colony was founded in 1630, and in the following decade English immigrants flooded in. The Puritan commonwealth was as intolerant as its foes, and in 1636 Roger Williams founded Providence in what later became the Rhode Island colony, where a wider religious liberty prevailed. Other immigrants also left Massachusetts, partly influenced by more material motives, to settle the richer lands of Connecticut.

In all the New England colonies there was a high level of participation in public affairs, creating an independent outlook that would even affect "proprietorial" colonies such as Maryland, which originated in 1634 as the personal property of Lord Baltimore. Emigration from England fell off during the 1640s, but by that time the colonies were a formidable presence on American soil.

HEREFORD

HEREFORDSHIRE described
With the true plat of the Citie Hereford, as also the
Armes of this Nobles that have bene intituled with
that Dignity

WORCESTER SHIRE

PART OF SHROPSHIRE

PART OF GLOCESTER SHIRE

NORTH

SOUTH

WEST

MOUNMOTH SHIRE

HEREFORDSHIRE
JOHN SPEED

In the sixteenth century the English counties were efficiently mapped for the first time. But even by 1600 towns, trade, and travel had developed so rapidly that an up-to-date county atlas was urgently needed—a need that was met by John Speed.

Speed's achievements were remarkable in view of his humble background. He was born at Farndon in Cheshire, probably around 1552, and, like his father, became a tailor. At some point he settled in London, where his passion for antiquities brought him into contact with a circle of aristocratic intellectuals. Thanks to their patronage, he became financially secure enough to study and write as he wished. Among the early results of Speed's labors were the maps he presented in 1598–1600 to Queen Elizabeth I and his own Merchant Taylors' Guild.

Though he was clearly fascinated by cartography, he probably believed that his most important work was his *History of Great Britain* (1611). The maps in *The Theatre of the Empire of Great Britain*, published in the same year, were primarily intended as a companion to the *History*, although some had been printed and sold individually a little earlier. Drawing heavily on the work of older English mapmakers, but updated, packed with textual and visual information, and extended to cover Scotland and Ireland, Speed's *Theatre* retained its authority for more than a century.

COUNTY TOWN
One of Speed's innovations was to include views of at least one important town in each map. Comparison with a modern plan shows that Hereford's basic layout has changed very little over the centuries.

WORKMANSHIP
The elegance and clarity of Speed's maps ensured their popularity. The credit belonged not only to Speed but also to the map's Dutch engraver, Jodocus Hondius.

THREE SUNS
The inset is an imaginary picture of the battle of Mortimer's Cross, Herefordshire, in 1461, during the Wars of the Roses. Three suns are said to have appeared in the sky before battle commenced.

COATS OF ARMS
Speed's maps display the coats of arms of the leading families in each county. They may well have sponsored the atlas to make sure they were included.

ENGLISH COUNTY MAPS

Maps such as Speed's Herefordshire belong to a type in which English cartographers excelled their otherwise more skilled contemporaries in continental Europe. The detailed surveying and mapping of the counties was completed during Speed's lifetime, establishing a distinctive English tradition.

Mapping with any real pretensions to accuracy began very late in Britain. Apart from the special group of medieval *mappaemundi*, only a handful of maps showing specific places seem to have been made. But in sixteenth-century England— the Tudor period—new ideas and influences entered the country from Renaissance Italy, and maps rapidly became familiar objects.

As a result, techniques improved rapidly. Topographical maps began to be based on scientific

Above: John Speed, pictured here in a contemporary engraving, was a renowned antiquarian as well as a cartographer.

surveying techniques such as triangulation, accurately transferred to paper by isometric drawing. By the end of the sixteenth century, conventional signs were being used for a range of features. As in other parts of Europe, however, the pictorial tradition persisted and splendidly decorated maps remained the rule.

Government patronage

Advances in the later sixteenth century owed a good deal to direct or indirect government encouragement. Laurence Nowell, who belonged to the household of the chief minister, William Cecil, made maps of the British Isles and Scotland that became part of Cecil's extensive collection. Nowell, an Anglo-Saxon scholar, also made a thirteen-section map of England and Wales with place names written in their Old English form and employing Old English letters—an eccentric procedure even by sixteenth-century standards.

In 1561 a clergyman, John Rudd, proposed to spend two years traveling in order to make a map of England. The government's interest in this kind of project is shown by the queen's order that Rudd should continue to receive his stipend as a prebendary of Durham while he was away. What became of the work is not known, and Rudd is mainly remembered as the "master" of a greater man, Christopher Saxton, who produced the first county atlas.

Saxton's benefactors

Saxton, a Yorkshireman, was born about 1542. In the early 1570s he found a patron in Thomas Seckford, who was surveyor of the Court of Wards and Liveries. Seckford's role was clearly important, since his coat of arms appears on every one of Saxton's county maps. However, Cecil's hand can also be seen in royal grants to Saxton and orders that localities were to give him every assistance in carrying out his surveys.

Between 1574 and 1578 Saxton issued thirty-four county maps and a general map of England and Wales, published together in 1579 as the first national atlas. This implies that in less than ten years, Saxton had surveyed as well as mapped the entire country— a feat so extraordinary that it suggests he must have made some use of Rudd's work or other now-lost maps. What makes Saxton's achievement even more remarkable is that his maps were extremely accurate as well as attractive.

Saxton's works were to be the basis of most county maps for the following century or more. He might have found a rival in the equally gifted John Norden, who planned a *Speculum Britanniae* ("Mirror of Britain") but managed to complete only nine maps of his projected county atlas; three of the maps were not even published during his lifetime. Though highly regarded, Norden seems to have been unable to secure adequate financing, perhaps a reflection of the difficulties of the 1590s. Among other improvements, Norden's maps included main roads for the first time. Another atlas project was left incomplete after twelve

maps were published in 1602–1603 by the herald (Rougedragon Pursuivant) William Smith.

John Speed's works were revised versions of Norden and Saxton, with many new features including the division of counties into hundreds. His *Theatre* was published in thirteen editions, the last around 1770! A "miniature Speed" pocket atlas was issued as early as 1627, with the maps engraved by Pieter van den Keere, and subsequently the great Dutch map publishers based their own English atlases on Saxton and Speed. The tradition they founded was continued in the eighteenth century, when exceptionally large-scale county maps were produced. These, in turn, paved the way for the Ordnance Survey (page 228).

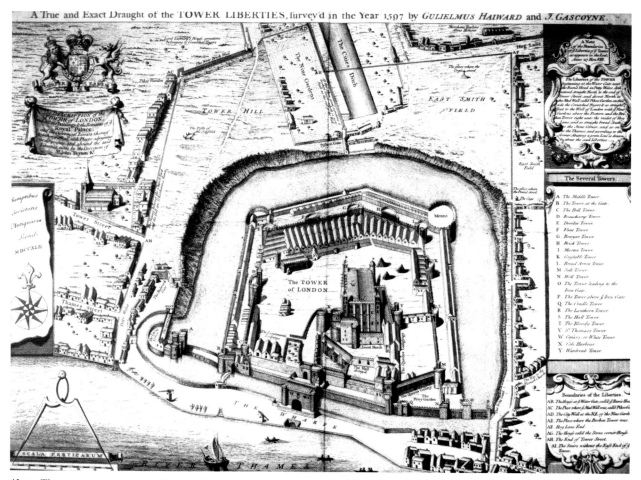

Above: The accurate surveying of England's countryside and rural towns found a parallel in the detailed description of its great cities. This bird's-eye view of the Tower of London and its environs was published in 1597.

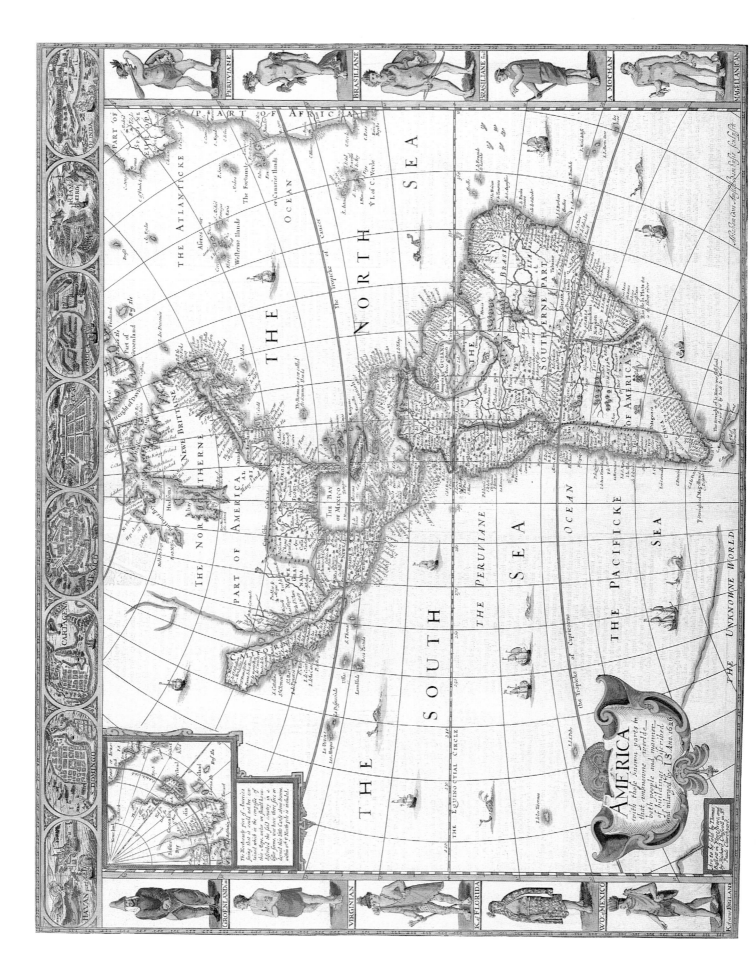

AMERICA with those known parts in that unknowne worlde both people and manner of buildings Discribed and inlarged by I.S. Ano. 1626.

Are to be Sold by Thomas Basset in Fleetstreet and Richard Chiswell in St. Pauls Churchyard

AMERICA
JOHN SPEED

John Speed was the first Englishman to make a more than local impact
on cartography. As well as his classic collection of British county and historical maps,
in old age he issued the first British world atlas, a work that was much
copied by his famous Dutch successors.

The great achievement of Speed's later years was *A Prospect of the Most Famous Parts of the World*, a twenty-one-map atlas published in 1627. Its famous map of North and South America incorporated the very latest English and Dutch information about the continent. Speed benefited from already being in close contact with his Dutch contemporaries: His earlier county maps were engraved by Jodocus Hondius, and America is the work of Abraham Goos. The map carries *cartes à figures*, multipaneled border pictures in a style later characteristic of Dutch maps. The *Prospect* combined its decorative features with an exceptional amount of information, with vignettes, plans, views, written panels, and accompanying text. Its up-to-dateness had one drawback: New (mis)information caused Speed to represent California as an island instead of a peninsula. In other respects his map is notable for its modesty: Unlike most cartographers of the time, he left undiscovered lands in a visibly tentative state, and was content to describe the far-southern regions simply as "the unknowne world."

INCA CITY
Cuzco, the capital of the Incas, still displays its original regular layout, though "re-founded" by the Spanish. All the cities shown in the top border were Spanish or Portuguese possessions.

INSET
This panel, letting the reader know why Greenland appears in a separate inset, outside the main map, is typical of Speed's compulsively explanatory style.

CALIFORNIA
Speed's was the first atlas in which California appeared as an island. Based on information on a captured Spanish chart, this error appeared on maps until late in the 18th century.

UNKNOWN TERRITORY
Speed has not attempted to complete the outlines of Tierra del Fuego or the recently discovered States Land (Staten Island), separated from it by Le Maire's Strait. Nevertheless Staten Island's size is greatly exaggerated.

INTO THE PACIFIC

West of the Americas on John Speed's map, the Pacific is shown as vast and almost empty. Two centuries of Portuguese, Spanish, and Dutch voyaging added a little to European knowledge of the Pacific's western margins, but the ocean did not begin to yield up its secrets until the mid-18th century.

Above: The strait between Patagonia and Tierra del Fuego provided a perilous gateway into the Pacific for Ferdinand Magellan, for whom it was later named.

Thanks to a combination of geographical and economic factors, the exploration of the Pacific was painfully slow. The ocean covered about a third of the Earth's surface, but although it contained thousands of islands, none of them (beyond the fringes of Asia and Australasia) was of any great size. European navigational science was not advanced enough to chart the islands accurately, and a number of them were discovered only to be lost again. In any case Europeans mainly regarded the Pacific as a barrier or a highway. What they cared about were the riches of East Asia and the East Indies (present-day Indonesia). With one or two exceptions, Pacific discoveries were accidents of passage.

The Portuguese, having reached the East via the Cape of Good Hope, had no great motive to explore the ocean much beyond the Indies. They controlled trade with the Spice Islands, shipped Chinese tea and porcelain from the island of Macao, and developed trade relations with Japan.

The arrival of the Europeans

Treaty provisions meant that the Spanish could only challenge the Portuguese monopoly by sailing west from the Americas. The first European to see the Pacific was indeed a Spaniard, Vasco Nuñez de Balboa, who in 1513 crossed the Isthmus of Panama and claimed the "Southern Ocean" that stretched before him for Spain. The first successful attempt to find a sea route to the Pacific came in 1520, when the Spanish expedition led by Ferdinand Magellan passed through the stormy strait between Patagonia and Tierra del Fuego and headed for the Spice Islands. The Spanish suffered horribly during

a ninety-eight-day voyage that revealed the immensity of the Pacific; they made landfall in the Marianas (Ladrones) and discovered the Philippines before sailing on through known waters to circumnavigate the globe.

Later Spanish expeditions reached some of the Caroline and Palau Islands, and in 1545 Ortiz de Retes landed briefly on, and named, New Guinea. Spanish activity increased after 1565, when Andres de Urdaneta discovered a route from the Philippines to Mexico, after which the settlement of the islands followed. In 1567 Alvaro de Mendaña discovered the Solomons, but when he returned in 1595 he could not find them again. Such were the confusions of sixteenth-century Pacific navigation that when he reached the Marquesas, some 3,100 miles (5,000 km) to the east, he believed he had reached his goal.

A similar air of futility hangs over the voyages of Pedro Fernandez de Quirós and Luis Vaez de

Torres. Quirós, who had been Mendaña's pilot, probably landed on Tahiti. When he reached the New Hebrides (Vanuatu) he became convinced that he had found the elusive southern continent and named the islands Terra Australis, confusing mapmakers for some time to come. Torres, one of Quirós's lieutenants, sailed south along the coast of New Guinea and found the straits now named after him. His proof that New Guinea was an island was never widely accepted and had to be repeated later by English explorer James Cook.

The Dutch dominate
During the sixteenth century the Portuguese and Spaniards remained unchallenged, despite some buccaneering exploits by Sir Francis Drake on his way around the world in 1578–1580. But in the 1600s the Dutch dislodged the Portuguese from most of their strongholds, and the Dutch East India Company became the dominant power in the region. Although Portuguese ships may have touched on Australia in the 1520s, the first certain reports of the

new continent were those of Dutch captains Willem Janszoon (1606), Dirk Hartog (1616), and Abel Tasman (1642). By 1663 Joan Blaeu's world map was showing a recognizable if incomplete "New Holland" (page 152).

The Dutch East India Company monopolized the route around the Cape of Good Hope, so in 1616 Willem Cornelisz van Schouten and Jacob Le Maire attempted to break into the spice trade by sailing west. They found a less hazardous route than Magellan's from the Atlantic into the Pacific, passing through the strait (Le Maire's Strait) between Tierra del Fuego and Staten Island and sailing around Cape Horn—very recent information that appears on Speed's map.

Europeans remained preoccupied with profit rather than knowledge, and as late as 1722 Jacob Roggeveen's discoveries of Easter Island and Samoa made no great impression. By the 1760s, however, there was a change of mood, and new nations entered and began at last to chart the immensities of the Pacific.

Above: The famous statues of Easter Island, today known as Rapa Nui, aroused little interest among profit-obsessed Europeans until the later eighteenth century, when exotic places began to attract scientific and popular curiosity.

LEO BELGICUS
JODOCUS HONDIUS

The outlines of maps have often been distorted or falsified for political and propagandist reasons. Less commonly, the map image is adapted to move or entertain. Among the most celebrated of such curiosities are the *Leo Belgicus* (Belgic Lion) maps, in which the Low Countries are cast in the patriotic form of a roused and roaring lion.

The first *Leo Belgicus*, printed in 1583, was the work of Michael Aitzinger. It was so popular that versions were published by most of the leading mapmakers in the Low Countries until the mid-seventeenth century. In the early versions the lion stood on its hind legs, with its paw raised, and faced east; later there were standing, south-facing figures (like the example shown here) and also seated lions, their relaxed pose apparently inspired by the long period of peace between 1609 and 1621.

The martial connotations of the lion were appropriate, since in 1568 the seventeen provinces of the Low Countries had begun a long war of independence against Spain; the alert, combative Leo shown here was the work of Jodocus Hondius, dated 1611. Seventeen-province lions continued to appear long after it had become clear that Spain would hold the south, but *Leo Hollandicus* maps were eventually made with only the seven northern (Dutch) provinces in lion form.

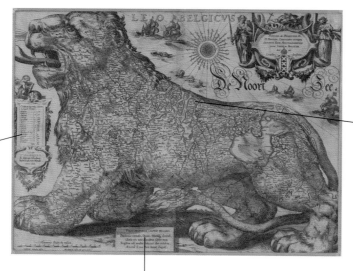

PROVINCIAL LIST
A framed list of the 17 provinces records the number of towns (*civitates*) and villages (*pagi*) in each. These reveal the striking extent of urbanization in the region.

PLEA FOR PEACE
Though Leo Belgicus looks ferocious, the panel makes him, as the personification of the Low Countries, quote the Roman poet Virgil on the supreme felicity of peace; a truce with Spain had been declared just two years earlier.

THE LION'S BACK
The lion shape is at its most convincing along the coastline, where it suggests an animal's back and tail. A line of "hair" serves to smooth out irregularities created by islands and inlets.

NOVELTY MAPS

Most maps have been designed to inform, edify, or give aesthetic pleasure. But sometimes the main purpose has been to surprise and entertain, either through ingenious visual effects or by placing the map in an unexpected setting. Such novelties have a surprisingly long history, although few were actually made.

The different versions of *Leo Belgicus* were not the earliest novelty maps. In 1537 Johann Putsch (Bucius) pictured Europe in the form of a woman, and about 1580 Heinrich Bünting designed two fanciful maps, the better known of which showed Asia as the mythical winged horse Pegasus. However, anthropomorphic maps were made even earlier, in the 1330s, by Opicinus de Canistris, an Italian cleric employed at the papal court in Avignon in France. Though his work combines elements of the

Above: The novelty map as nationalist celebration: Armed and ready to defend herself against the world, Britannia rests on her shield, which carries the design of the Union flag.

mappamundi and the portolan chart, its quirky "human geography" puts it very definitely in the category of a novelty. Among its striking images are a full-face bearded man representing the Iberian peninsula and a North Africa in the form of a woman, shown in profile, who gazes at him across the Strait of Gibraltar. Opicinus's drawings of the Italian peninsula as a boot are the earliest-known examples of this image.

Since the sixteenth century, many humorous or satirical maps have been published, inspired either by the shapes of countries or by national stereotypes such as the Russian bear. In Britain Thomas Rowlandson and James Gillray's ferocious caricatures featured maps, and maps also appeared in the more genial political satires of the Victorians. Among the "serio-comic" maps of Fred W. Rose was "Angling in Troubled Waters," a European map of 1898 in which bodies in a variety of postures were ingeniously fitted into the outlines of the various states, complete with contemporary references including the Dreyfus case in France and the assassination of the Austrian empress Elizabeth. The angling theme appeared in the fishing lines held by the main characters, one of them a John Bull, representing England, with a hooked crocodile labeled "Egypt."

Above: The novelty map as caricature and stereotype: Scotland appears as a dour-faced, wind-swept old man in a kilt in this map designed in the nineteenth century.

There have been many maps of purely imaginary lands—that is, deliberate fictional creations rather than places (for example, Eldorado) that were wrongly believed to exist. Such maps often accompanied a book, from Thomas More's *Utopia* (1516) onward. In the case of Robert Louis Stevenson's *Treasure Island* (1883), the map he drew for his stepson actually inspired him to go on and write the story. Perhaps the most influential literary maps of modern times have been those of Middle Earth, drawn by J. R. R. Tolkien to illustrate his *The Hobbit* (1937) and *The Lord of the Rings* (1954–1955), which inspired a host of imitations.

Mapping the course of love

Imaginary places also figured, with accompanying story, in the "Land of Love" maps that were popular in the eighteenth century. Oddly reminiscent of medieval allegories, these charted the course of

Above: The novelty map as ironic commentary: Denmark and its islands bear unflattering portraits of the royal family.

courtship through rocks or other hazards (representing lust, jealousy, or scandal) to marriage, or pictured the heart as a stronghold, complete with bastions and other contemporary fortifications that had to be stormed or captured by guile.

Maps appeared in new contexts from the late sixteenth century, when the English counties were represented on playing cards. In more recent times, both true and imaginary maps have been seen on items as varied as cigarette cases and T-shirts. Even more popular has been their use as the basis for board games, found as early as the mid-seventeenth century. Many were intended as geographical and moral teaching aids, but in the twentieth century entertainment became paramount. The classic Monopoly board can be regarded as a diagrammatic tour around London. And in Diplomacy and many of its war-gaming successors, historic or imagined conflicts are re-enacted on maps or representations of the original terrain.

Above: The map as nationalist propaganda: Here Italy takes on the appearance of Giuseppe Garibaldi, one of the leaders of the campaign for Italian unification, finally achieved in 1871.

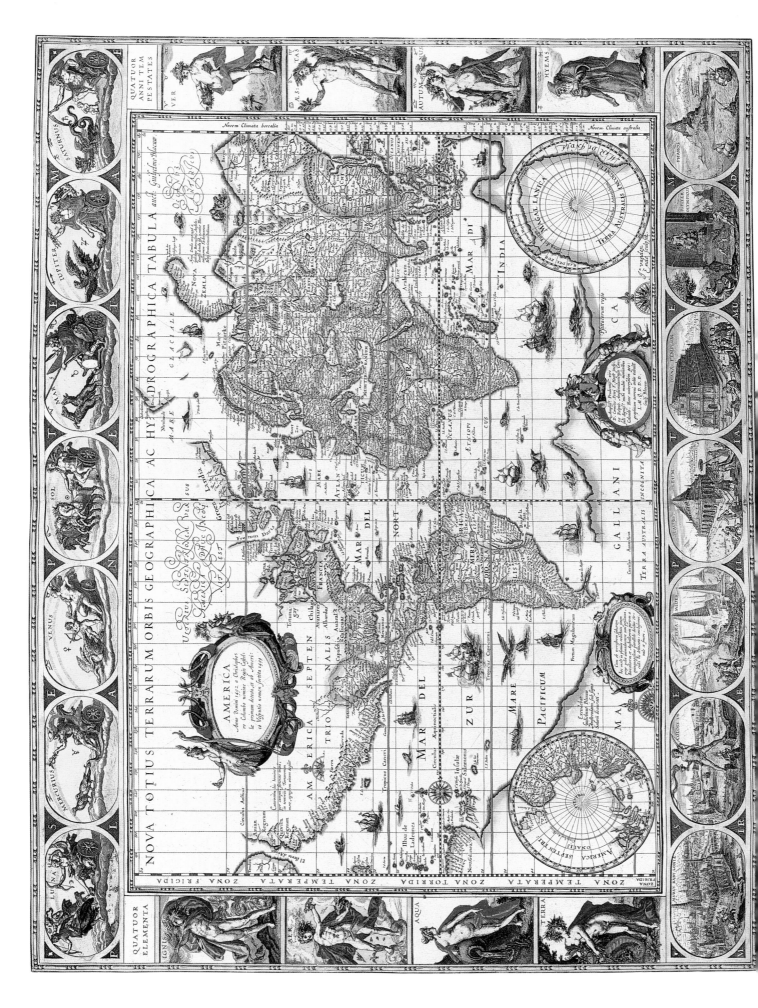

WORLD MAP
WILLEM BLAEU

One of the great names in the history of cartography, the Blaeu family were unequaled for ambition and decorative élan. Willem Blaeu's world map was a virtuoso effort, designed to dazzle clients in search of an atlas that would show off their wealth and good taste.

Willem Blaeu's world map was placed at the front of all the atlases issued by the Blaeus from 1630 onward. Only in 1662 did Willem Blaeu's son Joan replace it with the more up-to-date double-hemisphere version reproduced on page 152. Willem's showpiece map takes in everything: decorative, information-laden cartouches, sea pictures of ships and monsters, inset maps of the poles, and, above all, superb border panels filled with allegorical figures and scenes. The sun, moon, and planets are drawn through the sky by creatures linked to their divine personifications in classical mythology. On the left side are representations of the four elements (fire, air, water, earth); on the opposite side are the four seasons. The bottom strip displays the traditional Seven Wonders of the World. Everything is beautifully engraved and colored, with fine, clear lettering. A curious feature is the prominent cartouche celebrating the discovery and naming of America by, respectively, Christopher Columbus and Amerigo Vespucci.

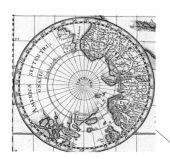

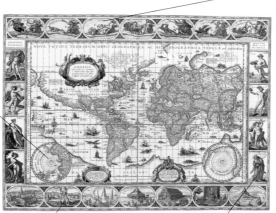

VENUS AND CUPID
Appropriately languorous, Venus is accompanied by her son, Cupid. He is blindfolded ("love is blind") and poised to fire a desire-arousing arrow. Venus's chariot is drawn by swans, birds that were associated with the goddess because of their beauty.

THE POLES
Inset maps of the North and South poles make attractive additions to the map. They are justified by the fact that the main map was drawn on Mercator's projection, which cannot show the poles.

COLOSSUS OF RHODES
The Colossus of Rhodes, dominating the ancient harbor. Other wonders are (left to right): Babylon, the Pyramids, the Mausoleum of Halicarnassus, the Temple of Diana, the statue of Jupiter (Zeus), and the Pharos (lighthouse) at Alexandria.

OLD MAN WINTER
Hyems, or *hiems*, is Latin for winter, shown here in traditional fashion as an old man (symbolizing "the dying year"), in contrast with the robustness of (top to bottom) spring, summer, and fall.

THE ATLAS MAKERS

Willem Blaeu's world map decorated a series of atlases issued by the Blaeu family, culminating in the *Grand Atlas*—the most expensive book published in the seventeenth century and still an unsurpassed achievement. In many respects, the work represents the zenith of the entire golden age of Dutch cartography.

Willem Janszoon Blaeu was born at Alkmaar in the Netherlands in 1571. Little is known of his early life, but as a young man he was well educated enough to be accepted as a student assistant by the famous astronomer Tycho Brahe, and he worked at Brahe's observatory on the Danish island of Hven. On his return to the Netherlands, he went into business at Alkmaar as a maker of terrestrial and celestial globes and instruments. By 1599, when his son and successor, Joan, was born, Willem had moved to Amsterdam. In the 1600s he printed and sold books; published maps of the Netherlands, Spain, and the world; and also issued sea charts and town plans.

The Blaeus prospered, and from 1605 the family occupied a more central premises in Amsterdam "at the sign of the gilded sundial." Until 1617 Blaeu signed himself Janssonius (the Latinized form of his second name), but as he became well known, his work was confused with that of a rival, Jan Jansson. The new signature he adopted was a Latin-Dutch mixture, G. or Gulielmus (Willem) Blaeu or Blaeuw.

Willem's first atlas

By the late 1620s Willem had published seventeen maps, but his stock was not extensive enough to enable him to issue an atlas. Then in 1629, following the death of the younger Jodocus Hondius, he bought between thirty-five and forty plates of maps used in the Hondius/Mercator atlases (that is, the posthumous editions of Mercator's atlas, as revised and expanded by the Hondius firm).

Willem rapidly brought out the sixty-map *Atlantis Appendix* (1630), the word "appendix" being used to associate the Blaeus' book with the Mercator/Hondius atlas and suggest that it was its natural, updated

Above: This illustration from 1587 shows the renowned astronomer Tycho Brahe in his observatory with a clerk, perhaps—who can tell?—mapmaker Willem Blaeu.

successor. In the following year an expanded edition of the *Appendix* appeared, with a preface by Willem boldly declaring that it completed the work of both Ortelius and Mercator. Published in four languages and containing about 208 maps, it was the first Blaeu atlas to carry the joint imprint of Willem and his son Joan. Since Joan was in his early thirties, while his father was a man of sixty (old by seventeenth-century standards), Joan may in fact have been running the firm for some time. However, about 1634 Willem was appointed cartographer to the Dutch East India Company, which effectively ran the nation's overseas commercial empire, and such a post would certainly not have been a sinecure.

The appointment may have been in part a recognition of the energy with which the Blaeus had issued maps, charts, and globes over the years. It may also have spurred them on to raise their reputation

still higher, especially since Henricus Hondius and Jan Jansson had entered into competition with them for the atlas market with a new edition of Mercator. At any rate, in 1635 the Blaeus produced the most ambitious of their publications to date, the *Novus Atlas* ("New Atlas") in two volumes. Depending on the language in which an edition appeared, it had 207 or 208 maps.

In 1636 Joan Blaeu laid the foundation stone for a larger premises on the Bloemgracht in Amsterdam. The Blaeus moved in the following year, and the Bloemgracht remained their headquarters to the end. Willem died in 1638, having laid the foundations for the future *Grand Atlas* that would give the Blaeu firm its cartographic immortality.

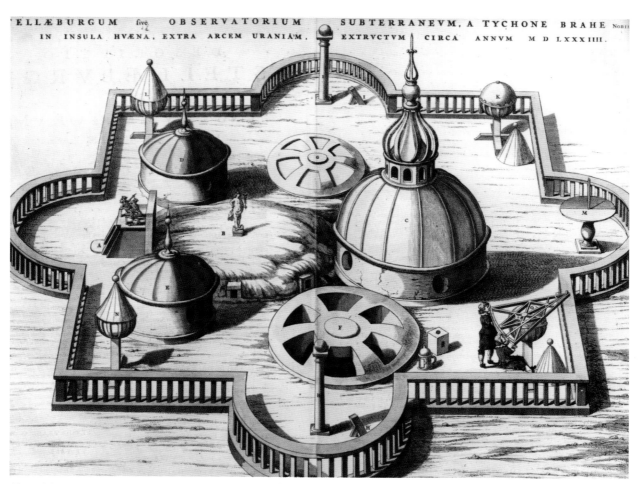

Above: A late-sixteenth-century view of the "Castle of the Heavens," the observatory built by Tycho Brahe at Uraniborg, on the Danish island of Hven, where Willem Blaeu worked as a young man.

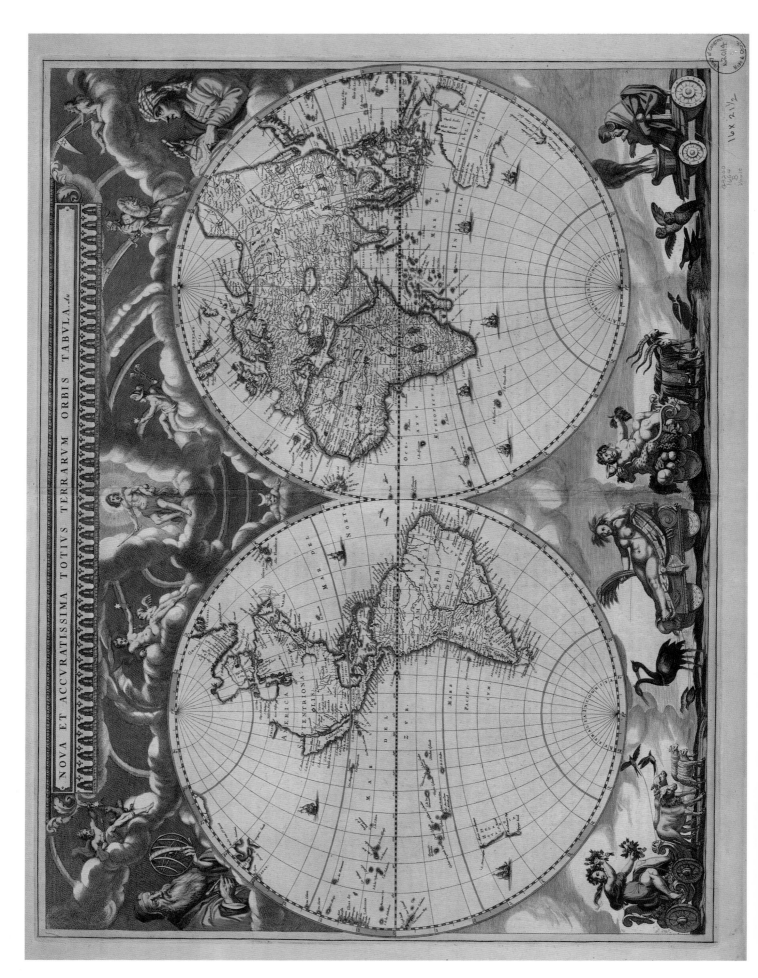

NOVA ET ACCVRATISSIMA TOTIVS TERRARVM ORBIS TABVLA.

WORLD MAP
JOAN BLAEU

From 1662 this double-hemisphere map of the world served as a showcase introduction to Joan Blaeu's multivolume *Grand Atlas,* the most ambitious and sumptuous map-publishing enterprise of the age. The new map was fully up to date both in content and in its decorative style.

Willem Blaeu's magnificent world map (page 148) prefaced the atlases issued by the family firm until it was replaced in 1662, twenty-four years after Willem's death, by the double-hemisphere map shown here. It was more in keeping with artistic fashion than Willem's map and more advanced in its content. The most striking change is the shrinking of the southern continent, which appears only as a geometric figure within the Antarctic Circle. New Guinea is shown as a separate land and so is New Holland, the future Australia. Joan has effectively admitted his (and other Europeans') ignorance by not completing the outlines of Australia and North America. However, a common contemporary error has been introduced in the form of an island California; and mythical Anian lingers on in the far northwest. Decoratively the map reflects the new Baroque style: The figures are freed from confining panels and are posed more dynamically, with cloud-riding gods above and the four seasons below.

SKY GOD
The planet Mercury is personified, like the other planets in the upper region of the map, by the god it was named after. Mercury, the messenger, is shown with his characteristic winged helmet and staff.

TYCHO BRAHE
The Danish astronomer Tycho Brahe uses a compass to take measurements on a globe. Brahe was the teacher of Joan Blaeu's father, Willem, and this imaginary portrait is consequently a tribute to both. Galileo works on his globe beside the other hemisphere.

FALL-TIME FRUITS
Fall is the time of abundance, suggested by the fruits. The tipsy-looking youth with a cup and grapes (and the leopardskin and goats, associated with the god Bacchus) is clearly celebrating the wine harvest.

AUSTRALIA TAKES SHAPE
"New Holland" takes shape at last. Its western and northern coastline is drawn with an accuracy that suggests a detailed knowledge of Dutch voyages of discovery.

THE BLAEU FIRM

Although Joan Blaeu replaced his father Willem's world map, his celebrated *Grand Atlas* was the end result of a long process involving both men. Joan's career was one of almost unbroken success until just before his death, when a catastrophic accident destroyed his entire concern.

Joan Blaeu was born at Alkmaar in the Netherlands in 1599, not long before the family moved and started a business in Amsterdam. He received a good education, and in 1620 was awarded a doctorate in law. His activities over the next few years are unknown; he may have worked with his father, although their names do not appear together until the 1631 edition of the *Atlantis Appendix*.

Joan takes over the firm

In 1636, following the publication of the two-volume *Novus Atlas* ("New Atlas") a year earlier, Joan laid the foundation stone for new premises on the Bloemgracht in Amsterdam. The family moved in the following year, and when Willem Blaeu died in 1638, Joan took his place as head of the firm and mapmaker to the Dutch East India Company. For five years he was in partnership with his brother Cornelis, who died in 1642. In 1640 the Blaeus issued an expanded, three-volume edition of the *Novus Atlas*. A year later, in August 1641, the English diarist John Evelyn visited the shop "to buy some maps, atlases and other works of that kind." He noted that "Mr Bleaw [sic], the setter forth of the Atlas's…is worthy seeing" but failed to specify what it was that made the publisher so striking.

In 1651 Joan Blaeu became an Amsterdam town councillor, the first map publisher to hold the office. Meanwhile he added more volumes to the *Novus Atlas*. Volume Four (1645) consisted of English county maps, mostly derived from John Speed's publications. Volume Five, issued in 1654, showed how widely Blaeu made contacts in his quest for new materials. It was the first-ever atlas of Scotland, its forty-nine maps based on surveys made by Robert Gordon of Straloch and Gordon's corrections of maps made around 1600 by a minister named Timothy Pont. (There were also six Speed-based Irish maps.) Finally, in 1655, Blaeu completed the *Novus Atlas* with a volume of maps covering the previously neglected Far East.

The *Grand Atlas*

The six volumes of the *Novus Atlas* provided the basic materials for an even more ambitious venture, the *Grand Atlas*. This first appeared in 1662 in an eleven-volume Latin edition, with almost 600 maps and 3,000 pages of text. The tall, splendidly bound volumes were luxury items, with fine calligraphy and lavish ornament, skillfully engraved on high-quality paper. Like all engravings, the maps were black

Above: John Evelyn, the English diarist, visited Joan Blaeu in Amsterdam. His diary hints that the cartographer himself was as memorable as his maps, but never enlarges on the hint.

and white, but whereas it was normal for the buyers to arrange for any hand coloring to be done, Blaeu's advertisements quoted two prices for the atlas, plain and colored, indicating that an in-house service was available. Since the *Grand Atlas* was expensive—probably the single most expensive publication of the century—most customers were probably sufficiently well off to take advantage of the specialist service offered, acquiring volumes that were richly but discreetly colored, with touches of gold as subtle reminders to anyone who looked through them that the owner was wealthy as well as tasteful.

The *Grand Atlas* proved popular: It was subsequently issued in French, Dutch, German, and Spanish editions. The Spanish edition was never completed, however, because many of the engraving plates were destroyed in a disastrous fire in February 1672. The entire Blaeu premises went up in flames, along with Blaeu's nine printing presses, type foundry, copper plates, books, and quantities of printed sheets and paper. Inevitably, Joan Blaeu's death the following year has been attributed to the severity of the blow. Any plates and books that had been saved were sold, and the great firm of Blaeu effectively ceased to exist.

Above: The Blaeu family chose a promising marketplace. Amsterdam was thriving and the Blaeus found ready customers among the rich merchants who were lining the canals with fine houses that were a quiet testament to their wealth and taste.

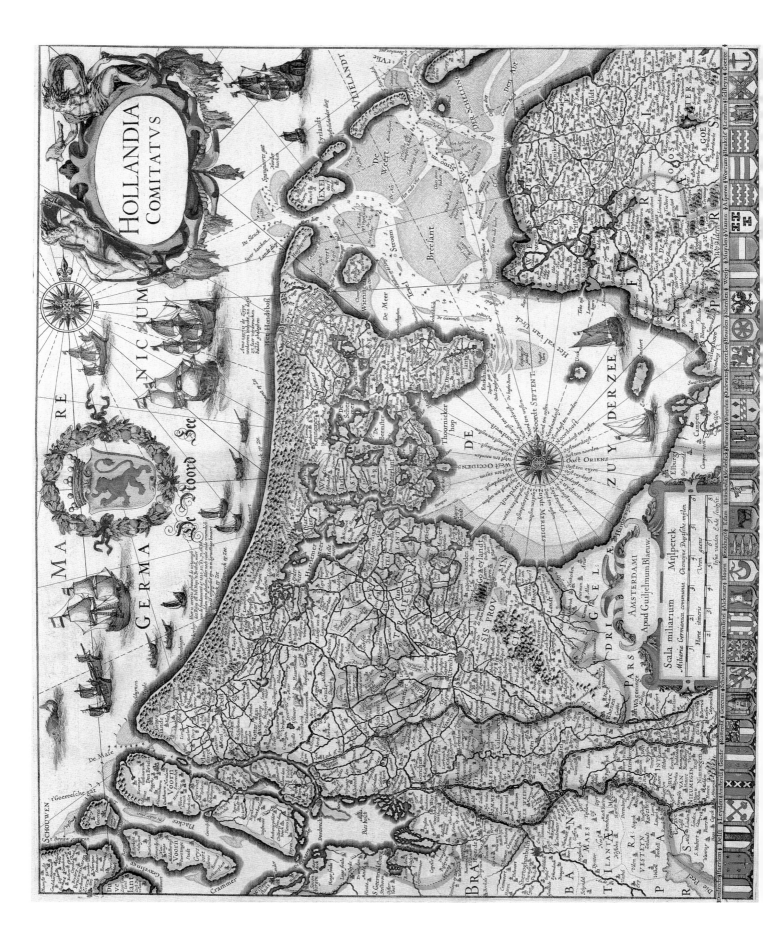

HOLLANDIA COMITATVS

Scala miliarium
Milliaria Germanica communia. Ghemeyne Duytsche mylen.
Horæ itineris
Amsterdami
Apud Guiljelmum Blaeuw.

Mijlperck

HOLLAND
WILLEM BLAEU

In the seventeenth century Holland was the most important of the seven provinces that made up the Netherlands. Blaeu's map shows it as it approached the height of its greatness, with picture messages stressing the vital part played by Dutch fisheries and command of the seas. Placing the North Sea at the top also emphasizes the role of seafaring and fishing.

The seven provinces had banded together in 1579 to resist the oppressive policies of their overlord, King Philip II of Spain. Effectively securing its independence by 1609, the new state became known as the United Provinces, United Netherlands, or Dutch Republic, as distinct from the ten southern provinces, or Spanish Netherlands, which became a nation state (Belgium) only in the nineteenth century. Willem Blaeu's map appeared in his 1630 *Atlas Appendix* and subsequent Blaeu atlases.

In 1662 it was one of sixty-five maps of the Low Countries published in the third volume of the *Grand Atlas*. Like its companions, Willem's map was made during the Dutch Golden Age, when Amsterdam was the financial heart of Europe, the Netherlands were the supreme maritime power, and the Dutch East India Company had a near-monopoly of trade with the East. Orienting the map with the west at the top underlines the importance of seafaring and fishing, as well as placing the Zuiderzee neatly in the center.

SHIPPING
Apart from a solitary concession to fashion in the form of a sea monster, the North Sea is shown crowded with Dutch vessels, from warships to small boats used in the shallows and on canals.

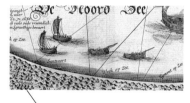

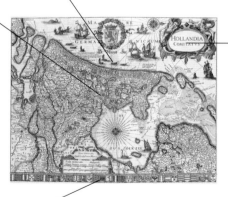

HOLLANDIA
The decorative elements around the title cartouche combine the classical and the mundane in an entertaining way. Above there are imposing sea deities; below are fish hanging up ready for market.

AMSTERDAM
Despite the shoals of the Zuiderzee, and thanks to a long inlet off it, Amsterdam could be reached from the North Sea and became a great commercial and maritime center.

WORKADAY
The coats of arms of most of Holland's towns are proudly assertive, flourishing lions, eagles, bars, and waves in the best heraldic tradition. A few, however, like this group, give a more direct account of themselves.

THE DUTCH GOLDEN AGE

Willem Blaeu's map of Holland shows a soggy, low-lying region whose ports could be approached only through shallows and sandbars. Yet this unpromising place nurtured a dynamic society of seafarers, merchants, manufacturers, financiers, artists, and scientists, who made the seventeenth century the Dutch Golden Age.

T he seven-province Dutch Republic was an accident of history. It was separated from the ten closely related "Flemish" provinces to the south by a military frontier that marked the farthest line of advance by Spanish forces in their partial reconquest of the Netherlands.

If anything distinguished the northern provinces, it was their paucity of natural resources and economic inferiority to the south, which was famous for its textile manufactures and great financial metropolis, Antwerp. Both south and north were unusual in the number and leading role of their cities. Having to feed their urban populations prompted the Dutch to develop more efficient cattle-raising and dairy industries than most contemporary societies, and to make the most of their opportunities for fishing. In their determination to harvest the ocean, the Dutch also became whalers, with their headquarters at distant Spitsbergen in the Arctic Ocean.

Fishing protected the Dutch from famine and also provided them with lucrative exports. This was important for trade with the Baltic, on which the Netherlands relied for grain, which it both consumed and sold on farther south. Even when Eastern markets had opened up, the Baltic remained the greatest single source of Dutch wealth, and from time to time the republic was drawn into regional conflicts to defend its interests.

The struggle against Spain stimulated Dutch enterprise, and in the 1590s the economy took off. Amsterdam became a boom city, benefiting from the decline of war-torn and Dutch-blockaded Antwerp. The Bank of Amsterdam was set up in 1602 and the Amsterdam Bourse (stock exchange) in 1611. Credit, exchange, and insurance facilities were developed on

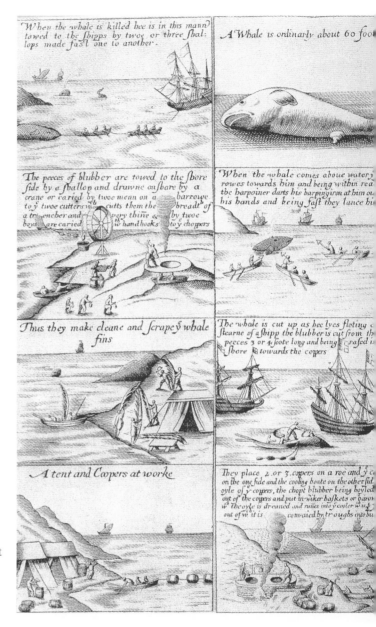

Above: One of the foundations of Dutch power was whaling, illustrated here in an American pamphlet from 1631. The Dutch based their operations in the Arctic Ocean.

a previously unseen scale, and Amsterdam became the center of the European money market.

Meanwhile in 1599 the first Dutch argosy had returned from the East Indies, loaded down with spices that brought in colossal profits. Within a few years Dutch traders were penetrating the Mediterranean and negotiating trade arrangements with the Ottoman Turks. Dutch shipyards built a new type of merchant vessel, the *fluyt*, that was capacious and cheap to run. It helped make the Dutch the carriers of Europe, able to undercut all their competitors' freight charges. They also built ships for their own and others' use, creating a major industry in spite of having to import the required timber.

The Dutch empire
Dutch vessels traded in distant parts of the world and, after initial doubts, laid the foundations of a colonial empire. New Amsterdam on Manhattan Island became the nucleus of a New Netherland in North America. Footholds were secured in the Caribbean and in Guiana, but Dutch designs on Brazil were frustrated. Following the establishment of the Dutch East India Company in 1602, the Portuguese were expelled from their strongholds in the Pacific, English attempts to compete were swept aside, and the Dutch, based at Batavia (Jakarta) in Java, monopolized trade from India to Japan. They settled the Cape of Good

Hope and, almost incidentally, were the first Europeans to reach Australia and New Zealand.

Dutch vigor was equally evident in urban society, its sobriety leavened by the arrival of Huguenots, Jews, and refugees from the south, who were attracted by the Netherlands' exceptional religious tolerance. A Dutch school of painting grew up that would have been distinguished even without Rembrandt and Vermeer, and great names in other fields included the philosopher Spinoza, the scientific pioneers Christian Huygens and Anton van Leeuwenhoek, and the jurist Hugo Grotius.

For a time the Dutch navy ruled the waves, while the army acquitted itself well against Spain and the even more formidable France of Louis XIV. The effort was exhausting, though, and in the long run the Dutch could not compete with larger powers. Overtaken politically and economically, in the eighteenth century the Netherlands entered a less exciting and creative, though still comfortably prosperous, era.

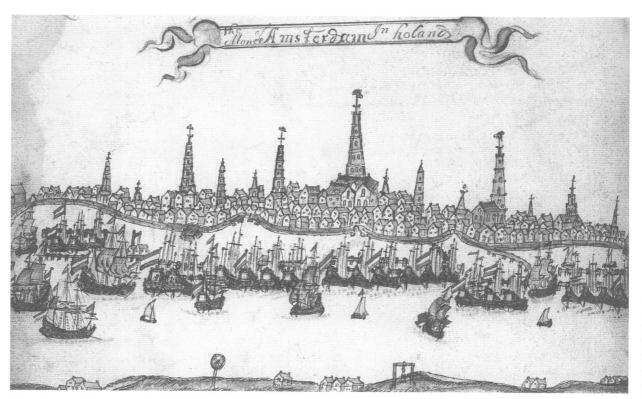

Above: A view of the busy harbor of Amsterdam, drawn by an English visitor around 1670. By that time Amsterdam's impressive churches and tall, ornate gabled houses had risen as proud symbols of the city's status as Europe's financial center.

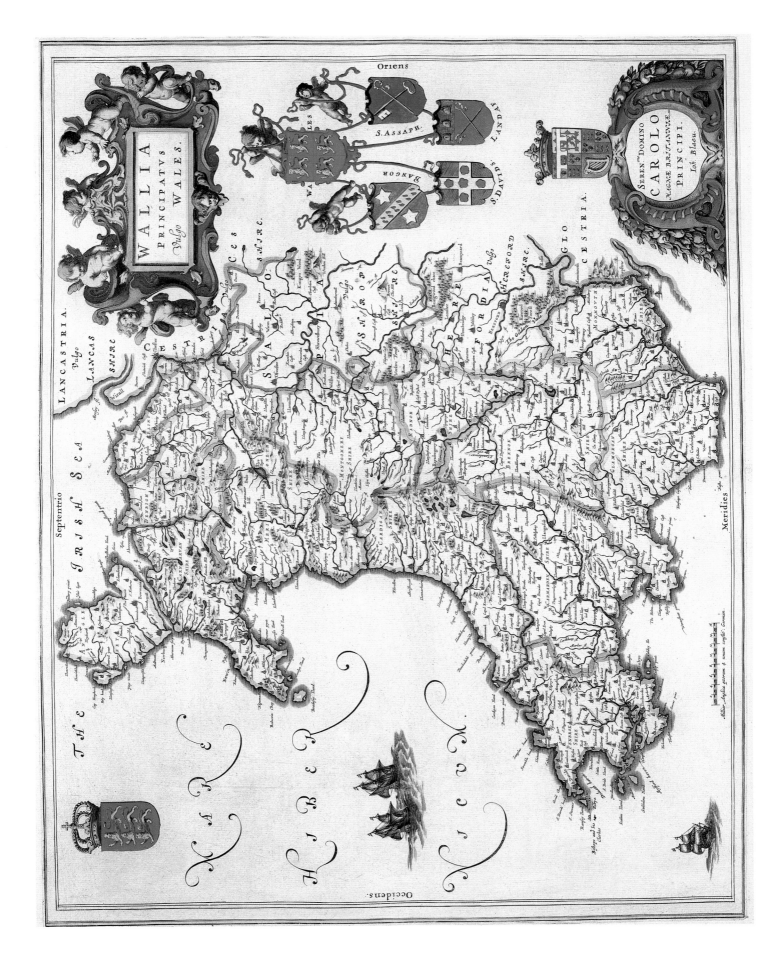

WALLIA PRINCIPATVS Vulgo WALES.

SEREN.mo DOMINO CAROLO MAGNÆ BRITANNIÆ, etc. PRINCIPI. Ioh. Blaeu.

Oriens

S. ASSAPH.

S. DAVID's

BANGOR

LANDAF

Septentrio

IRISH SEA

THE

ST. GEORGE'S

Meridies

Occidens

WALES
JOAN BLAEU

This fine, spacious map of Wales is one of Joan Blaeu's most visually satisfying productions, as well as being a model of accuracy. The decorative elements are much in evidence but are so skillfully arranged that they balance and enhance the purely cartographic features. Of particular note are the mini-landscape features and the episcopal arms borne aloft by cherubs.

Blaeu's map of Wales was first published in 1645 and later appeared in the fourth volume of the *Grand Atlas*, which contained fifty-eight maps of England and Wales. From a strictly cartographic point of view, Blaeu's account of Wales owed almost everything to the map in John Speed's *Theatrum* of 1611, but the freedom and decorative panache of the later map make it far superior. The map is dedicated to King Charles. (In 1645 the monarch in question would have been Charles I; by 1662, after the king's

execution and eleven years of a republic, it was Charles II.) Charles is described as ruler of Great Britain, a term popularized at the beginning of the century by James I and VI, the first king of both England and Scotland; but James's hopes of uniting his separate kingdoms into a single Great Britain had been frustrated, and an Anglo-Scottish union took place only in 1707. Wales had been formally incorporated into the Tudor state in 1536, but preserved its distinctive national identity.

CAERNARFON
Situated on the Menai Strait between the mainland and Anglesey, "Caernarvan" was said to have been the birthplace of Edward II, the first (English-created) prince of Wales. Surprisingly, the famous castles here and elsewhere along the coast are not mentioned.

HIGH PEAK
"Snowdon Hill" is now usually described more respectfully as Mount Snowdon. One of its five peaks is the highest in England and Wales.

TINY TOWN
Despite its Roman origin, Cardiff was still a small place on the Taff River when the map was drawn. Its big-city status dated from the 1840s, when it became a major port handling Welsh coal and iron.

EPISCOPAL CHERUB
This cherub is one of a group carrying beribboned shields bearing the arms of Wales and its four dioceses. St. Asaph's arms consist of a symbolic key and a bishop's crosier.

UNITING THE KINGDOMS

Blaeu's map of Wales appeared in the fourth volume of his *Grand Atlas*, devoted to England and Wales; Scotland and Ireland occupied a separate volume. Whether deliberately or by accident, the arrangement reflected the reality of Anglo-Welsh integration and the incompleteness of the process that would create the United Kingdom.

In 1500 the British Isles were inhabited by four peoples with a strong sense of nationhood and distinct linguistic traditions: the English, Welsh, Scots, and Irish. The larger population and resources of the English state made it likely that in the long run it would absorb its neighbors, but the process was long drawn out. During the Middle Ages the principal victim was Wales, though its conquest by Edward I in the 1290s necessitated building a ring of mighty castles. Even so, the Welsh were capable of staging a great popular revolt more than a century later under the leadership of Owain Glyn Dwr (Owen Glendower). English rule was maintained not by the Crown but by a group of nobles, the Marcher lords, who ruled vast border estates in semi-independence.

Campaigns of conquest

Elsewhere the English were less successful. In the twelfth century Henry II was acknowledged as "lord of Ireland," but control was difficult to maintain, and by the late fifteenth century, the royal writ ran only in the Pale, the area immediately round Dublin. Scotland remained completely independent. The most determined English effort at conquest, started under Edward I, ended in disaster for his son, Edward II, at the Battle of Bannockburn in 1314.

In Europe the sixteenth century was the great age of state-building, when strong, centralized monarchies began to eliminate internal competition. In England the Tudors were the state-building dynasty, and their long arm soon reached Wales (where their Welsh descent was probably an advantage). The power of the Marcher lords was broken under Henry VIII, and in 1536 an Act of Union made England and Wales a single country. Since the Welsh had been discriminated against ever since their revolt under Glyn Dwr, they benefited from having the same rights as the English. Wales could send Members of Parliament to Westminster, in London, and there were now many opportunities for gifted Welshmen to rise in the world—though the price was the adoption of English ways and speech. The Welsh language was frowned upon for centuries, surviving against the odds.

Henry VIII proclaimed himself king of Ireland in 1541, but effective authority became harder to exercise as England moved toward Protestantism, while the Irish remained fervently Catholic. The religious divide became even more important under Queen Elizabeth I, when Irish support for enemy

Above: Queen Anne, whose lack of children meant that England and Scotland might choose different monarchs after her death. The possibility led to the formal Act of Union in 1707.

Above: As imposing as when it was built by Edward I in 1283, Harlech Castle stands on a cliff overlooking the village of Harlech and distant Snowdonia. Despite its strength, the castle was briefly captured by Owain Glyn Dwr in the fifteenth century.

Spain prompted a major English effort and the first true conquest of Ireland, completed in 1603. In the same year the Scottish monarch, James VI, inherited the English throne as James I. Despite having the same king, the two countries remained separate—and mutually suspicious—states. James tried, but failed, to popularize the idea of a united "Great Britain."

During the turbulent seventeenth century, England and Scotland were sometimes at war, and Ireland was often in turmoil, further divided by Protestant settlement in the north and the creation of a Protestant landlord class elsewhere. For a time there was a forcibly united republic when Oliver Cromwell conquered Scotland and Ireland, but the status quo was reestablished in 1660 when the king was restored.

Even after the Protestant and constitutional settlement brought about by the "Glorious Revolution" of 1688, relations between England and Scotland remained problematic. The turning point was the Darien scheme, a Scottish colony established with high commercial hopes on the isthmus of Panama. Its failure almost bankrupted Scotland and made union with England (entailing access to English and colonial markets) much more attractive. Fortunately, English politicians felt the same, fearing that on the death of the reigning queen, the childless Anne, a disputed succession might enable the Scots to choose a monarch different from England's and become "the old enemy" again. With both political classes agreed, and despite a violent popular outcry in Scotland, the Act of Union was passed in 1707.

Great Britain proved a viable association, but its Irish policies failed to solve Ireland's ills—certainly not when, in 1800, the Irish parliament was heavily bribed to vote itself out of existence. This Act of Union created the United Kingdom, ruled from Westminster, but later events showed that, in important respects, "united" was a misnomer.

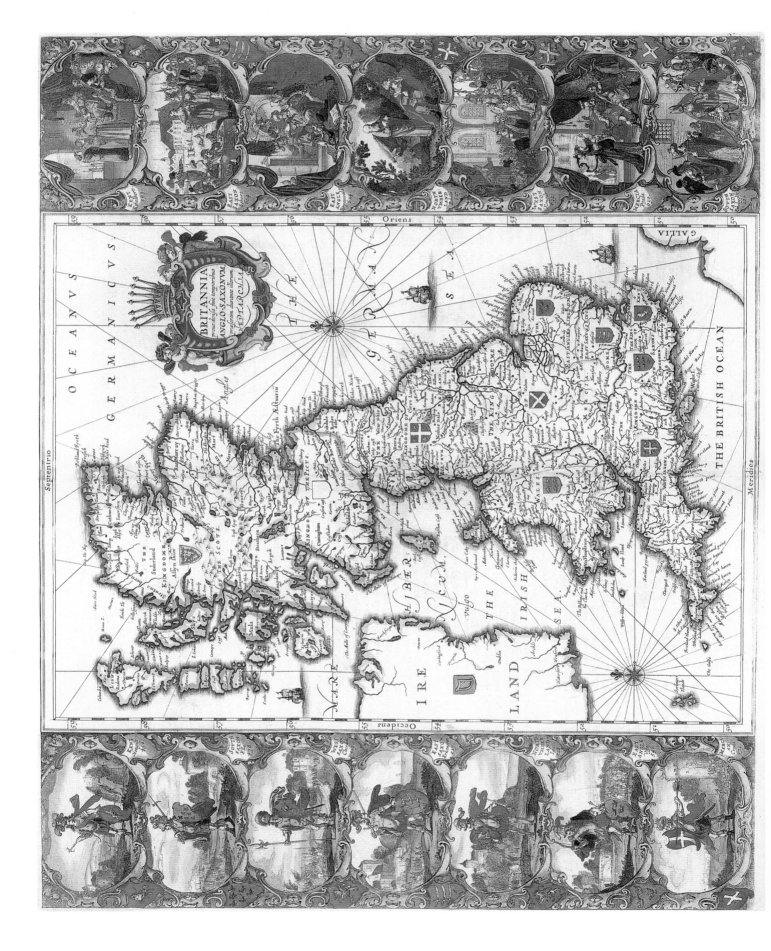

THE HEPTARCHY
JAN JANSSON

Interest in the Bible lands and classical antiquity ensured that maps frequently included historical references. By 1600 specifically historical maps were becoming more common in response to the development of Renaissance historiography and feelings of national identity. The Heptarchy, focusing on a relatively remote period of English history, is a notable example.

This is a historical map of Britain as it was in the sixth to eighth centuries, when England was divided into seven Anglo-Saxon kingdoms, often described as the Heptarchy ("rule of seven"): Kent, Sussex, Essex, East Anglia, Northumbria, Mercia, and Wessex. Jansson's map of 1646 is a near-identical copy of one by Joan Blaeu, issued a year earlier. Blaeu in turn had copied but greatly embellished John Speed's map of 1611, published as part of his British map collection, the *Theatrum*. England is shown as occupied by the Anglo-Saxons, along with the unconquered British ("Celtic") areas, Wales, Scotland (divided between the Picts and the more recently arrived Scots), and Ireland. On each side of the map are seven vignettes, each corresponding to one of the kingdoms. On the left, clad in the seventeenth century's operatic version of Roman costume, are imaginary portraits of each kingdom's founding father. Opposite are tableaux of incidents from the conversion of the Anglo-Saxons to Christianity.

PICTS
The kingdom of the Picts, who had once ruled all of Scotland. They were eventually absorbed by the Scots, leaving behind no significant historical records: hence the empty heraldic shield representing their kingdom.

CONVERSION
King Ethelbert of Kent receives Augustine, the head of a mission sent by Pope Gregory, which arrived in 597. Augustine's work in Kent began the conversion of England to Catholic Christianity.

FIRST SAXON KING
Cerdic was reputedly the first king of the West Saxons. He founded a dynasty that included King Alfred and ultimately united the Anglo-Saxons in a single kingdom.

VISION
Exiled and in fear of his life, the Northumbrian prince Edwin has a reassuring vision that will lead to his becoming king and effecting the conversion of Northumbria.

RECOVERING THE PAST

Jansson's map of the Anglo-Saxon Heptarchy, derived from John Speed's work of 1611, reflected profound changes in the mindset of post-medieval England. Interest in maps was one facet of a new, inquiring attitude toward history and an appreciation of England's topographical fabric, past and present.

Above: Elizabeth I ruled England from 1558 to 1603. Skillful foreign diplomacy, religious reform, and naval strength carried the nation through a difficult period in its history.

The Renaissance reached England in the sixteenth century, some time after it had taken root south of the Alps. It exercised a wide, though not always easily defined, influence on English society, which was just emerging from the Middle Ages. Curiosity about human achievement and human potential became much more intense, combined with a rising national fervor that became especially evident under Queen Elizabeth I.

One consequence was increased interest in people and places concerned with history and topography, as well as the making and use of maps. Generally speaking, medieval monastic chroniclers such as Matthew Paris and Ranulf Higden set their narratives within a religious or moral context, describing historical figures as exemplars of certain vices and virtues, and seeing the hand of providence in the way events developed. They tended to copy earlier works uncritically, often making the Creation their starting point. News or rumors that reached the monastery provided them with the materials they needed to bring their narratives up to date.

A new trend: antiquarianism

During the sixteenth century—the Tudor age— a greater interest developed in particulars, in documents, buildings, and other material remains. Its earliest manifestation was in a new kind of research, antiquarianism, rather than formal historical writing. Between 1535 and 1543 John Leland traveled up and down the land, carrying out a commission from Henry VIII to search for historical manuscripts in libraries—a useful exercise at a time when Henry was engaged in dissolving the monasteries. In the course of his duties, Leland made notes on everything he saw, from street names and building materials to ruined castles. He eventually went insane, perhaps despairing of his ability to put his material in order. Remaining in manuscript, his work was exploited by later antiquarians before achieving belated publication in the eighteenth century.

A few years later William Lambarde completed the first detailed study of a county, *A Perambulation of Kent.* As with maps, the government seems to have been closely involved in such projects. In 1573, while still in manuscript, it was sent to the chief minister, Lord Burghley, so that he could keep the queen well informed as she made an official journey through Kent.

By far the most influential of the Tudor antiquarians was William Camden, whose *Britannia* was published in the first of numerous editions in 1586. Originally intended as a study of Romano-British and Anglo-Saxon topography, the book was so amplified by Camden's incredibly industrious researches that it became a complete guide to Britain, only superseded when the nineteenth-century county histories appeared. Though written in Latin (still at that time the international language of scholarship), it was rapidly translated into English, and later editions included county maps based on those of Christopher Saxton and John Norden.

More specialized works continued to appear, notably John Stow's *Survey of London*, the first deeply researched account of any British town. Meanwhile the old chronicle tradition was carried on by Edward Hall and Raphael Holinshed, whose writings were plundered by Shakespeare for his plays. The playwright drew on Holinshed's stories of British history for a number of dramas, including the English history plays and parts of *Macbeth* and *King Lear*.

Early historical works

The first recognizable historical works in the modern sense were not published until the early seventeenth century. Sir Walter Raleigh, best known as a courtier and poet, was also the organizer of ill-fated colonial ventures and gold-seeking adventures in South America. A man of extraordinarily varied gifts, Raleigh wrote a *History of the World* (1614) when he fell from favor under James I and was confined to the Tower of London for thirteen years. His work was not a complete break with the past, but though he acknowledged the role of providence in history, he also sought for the mundane causes of events and displayed a lively sense of character.

Raleigh's work was enormously popular, partly thanks to its intense poetic spirit. This was absent from Camden's annals of the reign of Queen Elizabeth, published in 1615; but it nevertheless represented a significant advance, since it was based on state papers to which Camden's connection with Burghley gave him access. Then, seven years later, Francis Bacon finished his *History of the Reign of King Henry VII*. Famed for his advocacy of empirical science and systematic knowledge, Bacon also believed in history as a discipline that should "describe what men do, and not what they ought to do." These Jacobean works pioneered modern historical writing and paved the way for Clarendon, Edward Gibbon, David Hume, and other masters of the genre.

Above: Impressive timber frame buildings, such as Little Moreton Hall, are a physical reminder of the prosperity and solid comfort of English life during the Tudor age.

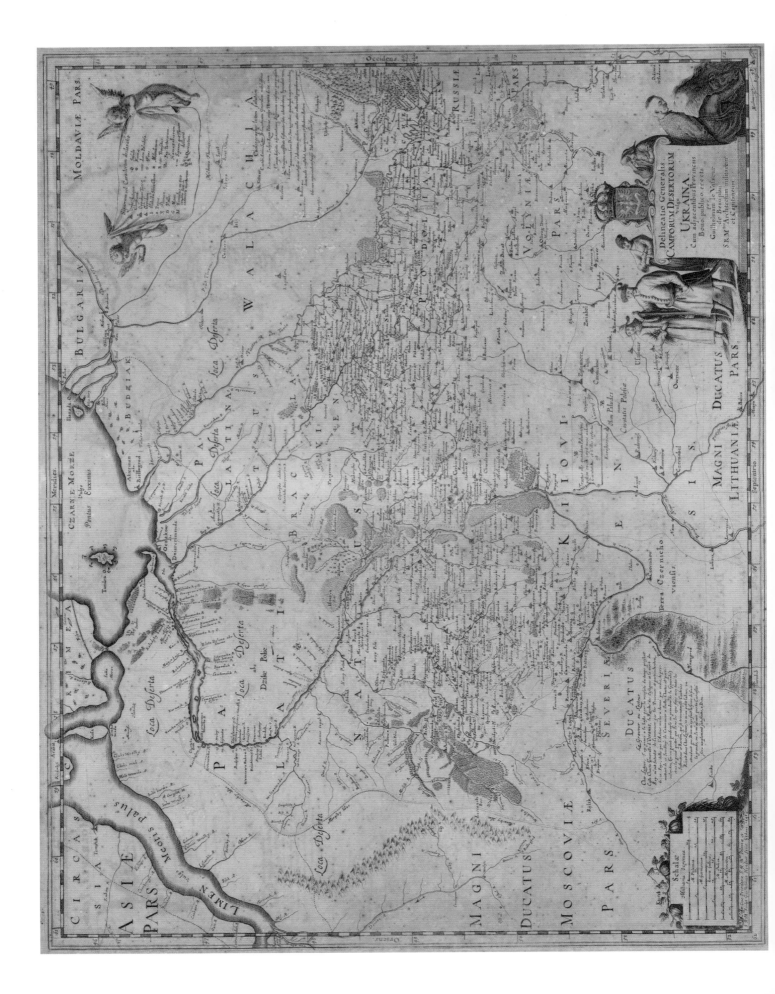

THE UKRAINE
GUILLAUME LE VASSEUR DE BEAUPLAN

Though mapping was being carried out in an increasingly scientific manner in the West, much of Eastern Europe was poorly recorded until experts arrived from abroad. One of them was the French engineer Le Vasseur de Beauplan, who made a number of maps of Poland before the country was plunged into crisis.

Le Vasseur was the son of a Norman pilot and chartmaker. He joined the Polish army in 1630, a time when Poland was a large and potentially powerful state whose kings were anxious to strengthen their authority by recruiting skilled foreigners. In 1639 Le Vasseur led an expedition down the Dniester River into the Ukraine, at that time a Polish province, and he subsequently made two maps including the one shown here. It was engraved and printed in 1648 by his friend Willem Hondius, a member of the famous Dutch family of cartographers who had settled at Gdansk. By that time Le Vasseur had left royal service, and other maps made by him and printed by Hondius were impounded because the information they contained was regarded as potentially useful to an enemy. On his return to France, Le Vasseur made a large map of Normandy and published a number of navigational works. In 1648 a Cossack revolt in the Ukraine initiated seventy years of civil wars and invasions that fatally weakened the Polish state.

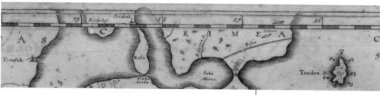

SOUTH
Unusually for this time, the map is oriented with south at the top. Hence the Crimea is shown with the Black Sea (Czarne Morze) and Bulgaria to its right.

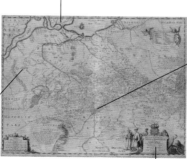

ANCIENT CAPITAL
Standing on the right bank of the Dnieper River, Kiev was the ancient capital of the Ukraine. Centuries before, it had been the center of the first Russian state.

ABANDONED
One of several places on the map marked "Abandoned Region." This one, scarcely populated since the Mongol invasions of 1238–1240, was to be colonized by free settlements (*slobodas*) of Cossacks, becoming known as the Sloboda Ukraine.

MIXED
A range of ethnic types gathers around the title and royal Polish coat of arms. Their presence reflects the turbulent history of the often-invaded Ukraine.

FRANCE ASCENDANT

Bringing their skills in engineering and war to Eastern Europe, experts such as Le Vasseur were living advertisements for the greatness of France. Under King Louis XIV, French arms dominated the Continent. Under his successors, French arts and sciences exercised an even wider, and ultimately subversive, influence.

I n 1648, when Le Vasseur's map of the Ukraine was published, the Thirty Years' War came to an end with Germany devastated and France on course to become the leading continental power. French supremacy was solidly based on the size, population, and resources of the country, and faltered only when distracted by internal conflicts or confronted with grand coalitions of enemies.

The conflicts were uppermost after 1648, when factions contended violently and royal power was eclipsed. Often humiliated during his minority, Louis XIV began his personal rule in 1661 determined not to be overshadowed by advisers or great nobles. Though served by brilliant ministers, he remained the master, making himself so central to the state that he became known as the Sun King.

As the setting for his solar presence, Louis built the vast palace of Versailles. Its furnishing—from tapestries and bronzes to silver and glassware—stimulated the decorative arts through official patronage on an unprecedented scale, from which the "Louis XIV style" emerged. Versailles also served political purposes, removing the court and nobility from volatile Paris and turning once-factious nobles into eager-to-please courtiers. Versailles became the model for the palaces of kings and princelings everywhere in Europe, and French classical norms predominated in art, architecture, music, literature, and drama, as exemplified by such famous writers as Racine, Corneille, and Molière.

Louis also dominated Europe in his pursuit of territorial expansion and military glory. In one campaign after another he asserted and enforced claims to areas in the Netherlands and the Rhineland. For a long time he managed to isolate his opponents before moving against them, but in time fear of his ambitions, and his growing religious fanaticism, united his enemies. The War of the League of Augsburg (1689–1697) and the War of the Spanish Succession (1701–1714) virtually bankrupted the country and brought Louis close to defeat. Even so, he was able to place his grandson Philip on the Spanish throne before he died in 1715.

Frivolity reigns

There followed an immediate reaction against the solemn grandeur and formality of Louis' court. The Sun King's great-grandson succeeded him as Louis XV. During his childhood, the Regency (1715–1723) was characterized by dissipation and frivolity, which were not much moderated when Louis himself began to rule. Louis himself is better remembered for his mistresses than his policies.

Above: Denis Diderot (1713–1784) exemplified the rational spirit of the Enlightenment. The seventeen text volumes and eleven volumes of plates of the great *Encyclopedia* represented an attempt to classify the whole of knowledge.

Meanwhile the obsolete features of the political, social, and economic system became increasingly obvious. France remained a great power, but its continental wars brought few advantages, while its colonial empire in Canada and India was lost to Britain. Ironically, during this period France did expand impressively in Europe, without firing a shot: the Genoese sold the troublesome island of Corsica to Louis XV, and a game of dynastic musical chairs left him in possession of Lorraine.

The Age of Reason

France retained its artistic and intellectual distinction throughout the period. The light-hearted rococo style of interior decoration provided a backdrop for the more informal society that developed, its members meeting in small rooms to experience the delights of good conversation, wit, and friendship. The most celebrated examples were the salons, presided over by intelligent women and attended by literary figures such as Voltaire, Jean d'Alembert, and Denis Diderot.

Eighteenth-century France demonstrated that wit could be directed to serious ends. Voltaire and most of his fellow writers were among the leaders of the Enlightenment, the great movement for rationality and against sterile traditions and obscurantism. They applied the test of reason to ideas and institutions and found many of them wanting. They were also enthusiasts for practical knowledge, and one of the great intellectual achievements of the age was the famous *Encyclopedia* edited by d'Alembert and Diderot, which disseminated a great deal of exact technical information as well as radical ideas.

A new era foreshadowed

The authorities made only sporadic attempts to suppress the *Encyclopedia*—partly through inefficiency and partly because the ruling class was itself infected by the new spirit. Aristocrats laughed at Voltaire's sallies but did little to repair the state or society. Within a few years of his death in 1778, the Age of Reason would give way to the Age of Revolution.

Above: The palace of Versailles, seen here from the Orangerie. It was intended by Louis XIV to be a physical manifestation of the power of the French state and the grandeur of his position at its heart.

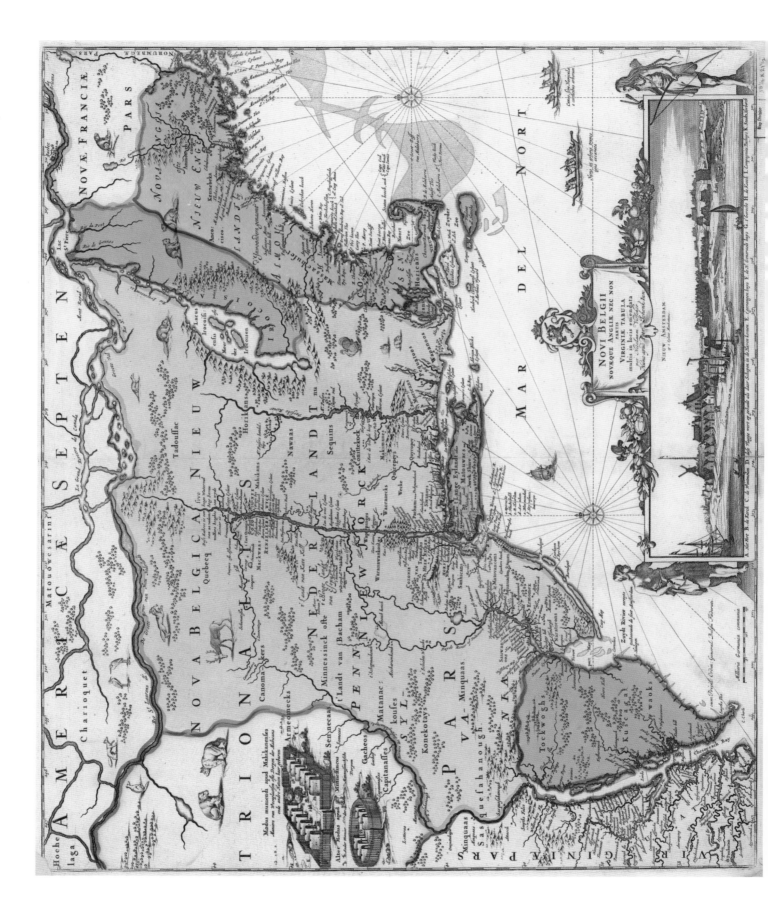

NEW NETHERLAND
NICHOLAS VISSCHER

New Netherland is a classic example of the way in which cartographers borrowed (or stole) from one another, embellishing if not always updating the material. Nicholas Visscher was only the second of a series of mapmakers and publishers who exploited this elegantly designed representation of the northeast American coast.

The long history of this map begins with its publication in 1651 by the firm of Jan Jansson. It showed the east coast of America from the Chesapeake to Maine, patriotically allocating most of the territory to the Dutch so that New Netherland dwarfed neighboring New England. Clearly engraved, with woods and mountains shown and sketches of Native American life and fauna, the map was both attractive and informative. The same was true of Nicholas Visscher's version, for the very good reason that when redrawing and engraving it in 1655, he faithfully copied it, slightly varying the fauna (adding the bears) and putting in a few more place names. His greatest single contribution was decorative: the vista of New Amsterdam (New York) flanked by Native American figures. A splendid new version was issued by Hugo Allard in 1673 to celebrate the recapture of the colony from the English—so splendid that it was reissued for decades, long after the final cession of New Netherland to England in 1674.

STRONGHOLD
Some Native American tribes built stockades like those shown on the map. The rectangular style was the one adopted by the Mahicans, or Mohicans, a tribe that became particularly well known thanks to Fenimore Cooper's famous novel of the same name.

TURKEY
The turkey was a native American species that was introduced to Europe in the sixteenth century. Other vignettes show amiable beavers, as well as foxes, bears, rabbits, and other creatures.

MANHATTAN
New Amsterdam is already named on the map, along with the Manhattans, who gave their name to the island. Long Island and Staten Island are shown and labeled just below it.

SETTLEMENT
New Amsterdam, soon to become New York. The settlement is small enough to include the windmill, churches, the fort, and so on. The presence of a gallows strikes a jarring note in an otherwise peaceful scene.

NEW NETHERLAND

The area shown on Visscher's map as New Netherland represented a generous and not entirely realistic account of Dutch territories in North America. By 1655 a failure to settle in sufficient numbers was weakening the Dutch position, and within a few years New Netherland and New Amsterdam ceased to exist.

Dutch claims on the North American mainland were based on the exploits of an English mariner, Henry Hudson. Already experienced in searching northern waters for a passage to the East, Hudson was commissioned by the Dutch East India Company to make another attempt by sailing round the top of Novaya Zemlya.

In April 1609 he left Amsterdam in a small vessel, the *Half Moon*, in which he rounded North Cape. The weather was so severe that his Anglo-Dutch crew mutinied and refused to go on. Rather than abandon the voyage, Hudson persuaded the crew to adopt a new course by showing them maps and letters sent by John Smith of Virginia. These suggested that the desired passage to the East could in fact be found on the North American coast.

The *Half Moon* reached Newfoundland in July and then sailed south as far as present-day North Carolina. Turning north again, Hudson reached New York's Lower Bay and hesitated for some time before passing through the Narrows and into the river that now bears his name. He had ventured farther than the only earlier explorer of the region, Giovanni da Verrazano, in 1524. He went some 150 miles (240 km) upriver, beyond the site of present-day Albany, before concluding that this was not the way to the Orient. During the journey he had traded profitably with the Native Americans. As so often in the history of exploration, the quest for mythical routes and riches produced unexpected benefits of quite a different kind, which the Dutch were quick to realize.

Dutch traders and colonists

Hudson himself, having returned to Dartmouth, was forbidden to continue serving a foreign company. He is presumed to have perished the following year, in the Canadian bay named after him, when mutineers cast him adrift in an open boat. Meanwhile, Dutch

Above: Imagining a historic moment: The painting celebrates the contacts made between Native Americans and the English navigator Henry Hudson, who in 1609 sailed up the river that now bears his name as far as modern Albany.

Above: This illustration, which dates from 1651, is said to be the first view of New Amsterdam, which at the time was little more than a fort with a few houses and a typically Dutch windmill.

merchants lost no time in trading for furs in the Hudson Valley, and in 1614 one of them, Adriaen Block, explored the coast to the northeast and sailed up the Connecticut River. However, trade rather than settlement remained the prime objective. The Dutch West India Company was established in 1621, but it was three years before a number of families were sent to settle at Fort Orange (Albany). They were followed in 1625 by the first colonists on Manhattan Island, who founded New Amsterdam.

In 1626 Peter Minuit bought Manhattan Island from the Native Americans. Though the fur trade remained lucrative, only a trickle of immigrants arrived from the Netherlands, and despite their claims, the Dutch controlled little territory beyond the Hudson Valley. In 1633 they built a fort on the Connecticut River at what is now Hartford, but English colonists from Plymouth settled upriver, and soon waves of new arrivals, there and on Long Island, greatly outnumbered the Dutch. In 1654, acknowledging realities, the Dutch abandoned their

post on the Connecticut River. By the time Visscher's map appeared, it was already out of date.

By the 1660s the Dutch population of New Netherland was still only about 10,000, whereas there were some 70,000 British settlers in North America. In New Amsterdam, governed by the unpopular Peter Stuyvesant, the Dutch constituted a bare majority. The demographic imbalance made them highly vulnerable to an English attack, which came not from the colonists but from outside. In 1664 Charles II of England granted all the lands between Connecticut and the Delaware—effectively all of New Netherland—to his brother James, duke of York. James sent a fleet that entered New Amsterdam's harbor, and when the settlers refused to fight, Stuyvesant surrendered.

New Amsterdam was renamed New York, and New Netherland was broken up into four colonies. Most of the Dutch inhabitants, offered easy terms, stayed in the Hudson Valley, maintaining their customs until long afterward and making a distinctive contribution to American life.

THE STARRY HEAVENS
ANDREAS CELLARIUS

Compilers of celestial atlases customarily enlivened their charts with traditional images
of the constellations derived from the pagan mythology of ancient Greece and Rome. The
results were of varying artistic worth, but the great work by Andreas Cellarius was a
spectacular masterpiece of the Baroque style.

Almost nothing is known of Cellarius beyond
what he chose to tell the world in the
introduction to his *Atlas Coelestis seu Harmonia
Macrocosmica*, first published in Amsterdam by Jan
Jansson in 1660: that he was the rector of a Latin
school at Hoorn in Holland. He may have been a
Dutchman, or possibly a German. He later published
a guide to the ancient world (1697). The atlas was a
comprehensive work, illustrating the stars newly
discovered in the southern sky by explorers during
the previous two centuries, and also constellations
recently seen for the first time with the telescope.
Newcomers are incorporated in the chart shown here,
along with traditional images including the Zodiac
constellations. The celestial sphere is seen from
outside, with the Earth at its center. An important
aspect of Cellarius's enterprise was an attempt to
provide Christian alternatives to the traditional pagan
names and images; though an artistic success, the
innovation never found wide acceptance.

ZODIAC
Cancer, or the Crab, one of the
twelve constellations of the
Zodiac which lie along the
ecliptic, the imaginary line
representing the passage of the
Sun through the heavens.

BIG BEAR
Ursa Maior, the Great Bear, has been
a familiar feature of the night sky
since antiquity. Seven of its stars form
"the Big Dipper" (the "Plough" in
Britain), pointing to the Pole Star.

NEW STARS
The constellation of Camelopardus
(an old name for the giraffe) was a
recent discovery thanks to the
invention of the telescope; it was
observed and named by Jacobus
Barschius in 1614.

OBSERVER
A youthful astronomer
with his cross-staff, a simple
but quite effective device
for measuring the height
of celestial objects. It was
also of direct practical
use in enabling mariners
to fix their positions
(latitude, but not yet
longitude) at sea.

CHARTING THE HEAVENS

The celestial atlas of Cellarius, with its pagan and Christian images, is a relatively recent example of the persistent human impulse to observe the skies and interpret them in terms of myth or symbol. By Cellarius's time, however, the scientific approach, freed from all preconceptions, was beginning to prevail.

The motions of the Sun, Moon, and stars fascinated prehistoric people and became the basis for mythical beings and events in most known cultures. The Western tradition, in which the constellations are linked with figures from classical myths, such as Pegasus and Castor and Pollux, is only one of many.

The alignment of prehistoric monuments so often corresponds with celestial events that, although detailed interpretations may be disputed, it is widely accepted that the builders possessed some astronomical knowledge and deliberately introduced it into their works and rituals. Such knowledge could be put to practical use, and the ancient Egyptians knew that when Sirius, the brightest star in the sky, rose above the eastern horizon at dawn, the Nile was about to rise and inundate—and fertilize—the land. Other cultures from Polynesia to Africa have also used the behavior of stars and planets to regulate their calendars.

The Egyptians introduced the 365-day year, and the Babylonians the units of sixty (minutes, seconds) and 360 (degrees). From at least the eighteenth century B.C., the Babylonians kept careful records of celestial phenomena that have survived, inscribed on clay tablets in cuneiform (wedge-shaped) characters. Although they eventually developed sophisticated mathematical applications, knowledge of the heavens was pursued mainly for its supposed use in foretelling the course of earthly affairs. Astrology and astronomy were similarly entwined in China, where the first star catalogs were compiled in the fourth century B.C. The oldest surviving Chinese chart dates from about A.D. 800 and shows more than 1,350 stars.

The Greeks try to measure the universe

Babylonian astronomy had a profound influence on the ancient Greeks, who inherited (and have passed on) names of constellations, such as the Bull, Lion, and Scorpion, that were already ancient when the Babylonian priest-observers employed them.

The Greeks themselves made the first thoroughgoing attempts to understand and measure the universe, in so far as that could be done by observations made with the naked eye and the power of thought. In the third century A.D., Aristarchus of Samos proposed that the Earth traveled around the Sun, and Eratosthenes calculated the circumference of the Earth with impressive near-accuracy. In the following century, Hipparchus's many important discoveries included the precession of the equinox. Greek astronomers made (flat) star charts and also the first celestial globes, based on what the heavens, conceived of as a sphere surrounding the Earth, would look like if viewed from outside rather than by

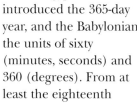

Above: This celestial globe, designed by Tycho Brahe in about 1590, was around 5 ft. (1.6 m) in diameter. Brahe was able to build magnificent instruments thanks to lavish royal patronage.

Above: This image of the constellations of Perseus and Andromeda from a fourteenth-century manuscript reflects a renewed interest in astronomy in the late Middle Ages.

looking up into the night sky. Greek and Roman charts and globes have all disappeared, with the single exception of the *Farnese Atlas* (c. 200 B.C.), a marble statue of Atlas holding up a celestial globe covered with symbols of the constellations.

Most of what we know about Greek astronomy comes from Ptolemy's *Almagest*, which contains his great star catalog and an influential exposition of the geocentric system. The authority of Plato and Aristotle had discredited Aristarchus, and Ptolemy set himself to explain anomalies in the behavior of the planets as they (supposedly) revolved around the Earth. His ingenious theory of epicycles (roughly speaking, planetary motion as circles-within-circles) dominated Western thought for a thousand years.

Rival views of the universe

For much of that time Ptolemy was unknown in the medieval West but preserved and valued in the Islamic world. There a number of Arab scholars made superb globes and astrolabes and compiled astronomical tables that were used in Europe for centuries. Ptolemy's *Almagest* reached the West via Sicily in the twelfth century, but scholars were slow to make use of his valuable methods and mathematics.

His geocentric universe, however, became dogma, not challenged until the sixteenth century, when Aristarchus's heliocentric theory was revived by Nicholas Copernicus. A compromise between the Ptolemaic and Copernican systems was proposed by the Danish astronomer Tycho Brahe, who was celebrated for his unprecedentedly accurate

instruments and observations. (Joan Blaeu's *Grand Atlas* contains eleven views of Brahe's observatory on Hven, where Willem Blaeu worked as a young man.) As is well known, Galileo championed Copernicus and suffered for it at the hands of the Inquisition, and even in 1660 Cellarius hedged his bets by illustrating all three systems in the *Atlas Coelestis*. However, he included the moons of Jupiter, first observed in 1609 by Galileo with an instrument that would revolutionize the charting of the heavens—the telescope.

Above: Tycho Brahe, the great Danish astronomer, pictured by Cellarius in a corner detail from his *Atlas Coelestis*. Brahe built a state-of-the-art observatory on the island of Hven in 1576.

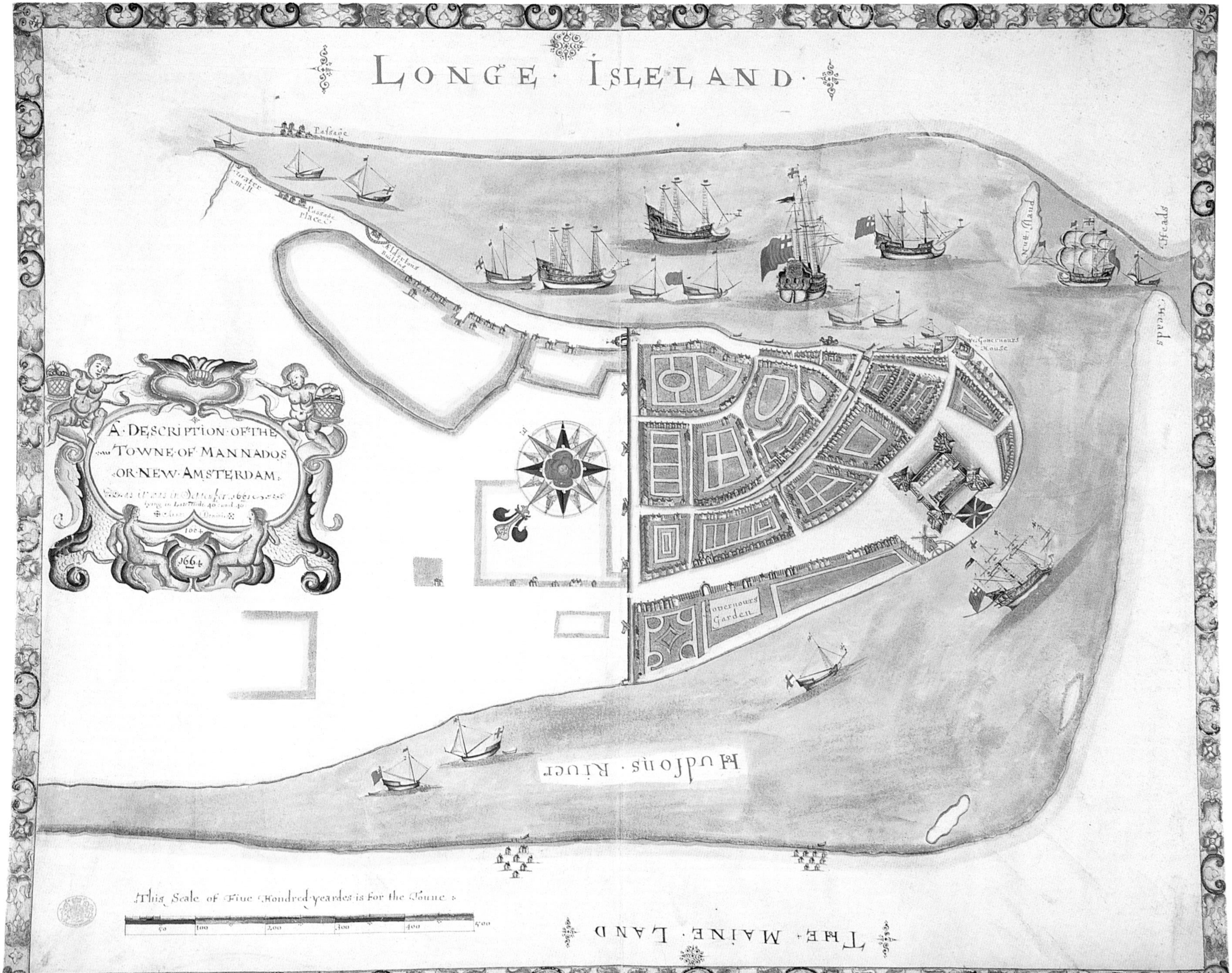

Heads

Nus Island

LONGE · ISLELAND ·

Gouernours Garden

THE · MAINE · LAND ·

Hudson's Riuer

E

Ellis place

Massart place

Burger Mill

A · DESCRIPTION · OF · THE
TOWNE · OF · MANNADOS
OR · NEW · AMSTERDAM

1661

This Scale of Fiue Houndred yeardes is for the Towne

TOWN OF MANHATTAN
NEW AMSTERDAM

The first English-language map of New York City was a pretty, anonymous chart that makes a pleasantly naive impression. Though probably based on a Dutch original, it celebrates the English occupation of the town, an event of even more historical significance than jubilant contemporaries realized.

Manhattan, or New Amsterdam, became a prize in the struggle for maritime mastery between the Dutch and the English. The English took the island in 1664 and renamed it New York, in honor of James, Duke of York, later James II, who was invested with the Dutch territories. The Dutch recaptured it briefly in 1673, but thereafter it remained in English hands until the American Revolutionary War. The map, "A Description of the Towne of Mannados or New Amsterdam," is something of a puzzle. "Mannados" is a Dutch/English version of "Manhattan," and the 1664 date, the ships, and other details make it clear that the town is in English hands. However, the fact that it claims to show New Amsterdam only "as it was in September 1661"—three years earlier—strongly suggests that the nondecorative elements were copied from a now-lost Dutch original. Little is known about the map in its existing form except that it was presented to the Duke of York and so is often referred to as "The Duke's Plan."

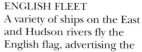

ENGLISH FLEET
A variety of ships on the East and Hudson rivers fly the English flag, advertising the role of the fleet in capturing New Amsterdam and the commercial benefits that would result from it.

CURLICUED CARTOUCHE
The crude but prettily decorated cartouche boasts floral and vaguely marine curlicues and full-fleshed attendant cherubs and nymphs, surrounding a plain yet puzzling inscription.

DIFFICULT NARROWS
The passage between Long Island and the New Jersey coast, though difficult to negotiate (especially given the location of what is now Governor's Island), is not as narrow as the map implies.

IMPOSING FORTRESS
The Dutch fort, with its four great bastions, looks formidable. However, when the English fleet sailed into the harbor in 1664, the Dutch surrendered without a struggle.

THE AMERICAN COLONIES

The anonymous 1664 map of Manhattan commemorated the taking of a small settlement and its sparsely populated hinterland. But the English victory had momentous consequences, creating a solid east-coast colonial bloc that, for all its diversity, had a potential for growth and a tradition of self-government.

By the 1630s English settlers were established in the colonies of Massachusetts and Virginia. Already a move had begun, impelled by religious persecution or land hunger, to expand settlement into Connecticut, Rhode Island, and Maryland. New immigrants continued to arrive until 1641, when the English became absorbed by the events of the Civil War and the radical changes that were taking place in their homeland.

The American colonies were largely left to themselves, even after the monarchy was restored in 1660, except for the expedition that seized New Netherland in 1664 and so created the "Middle Colonies" of British America. In the years that followed, New York developed only slowly, since settlers were not strongly attracted to the semifeudal system of large estates that had been taken over from the Dutch. By contrast, the land between the Hudson and the Delaware—East and West Jersey—was rapidly taken up. Prominent among the immigrants were Quakers, who became even more welcome farther inland following the grant of Pennsylvania to the Quaker William Penn by Charles II in 1681. A year later the city of Philadelphia, destined to become the largest in colonial America, was laid out. Meanwhile, the Carolinas were being settled, and in 1680 the ten-year-old township of Charles Town (Charleston) was relocated to its present site, Oyster Point.

Above: William Penn negotiates a land treaty with the Delaware. Penn advocated dealing fairly with Native Americans, for example, forbidding selling them alcohol to which they had no resistance. His policies actually attracted several tribes to move to the colony.

Above: William Penn, an English Quaker, founded Pennsylvania on land granted to him by King Charles II in settlement of a debt owed to Penn's father. He laid out Philadelphia—"the city of brotherly love"—in what became American cities' characteristic series of orderly rectangular blocks.

Rapid colonial expansion

Almost everywhere the colonists flourished. Their numbers rose rapidly, and by 1700 the population stood at about 250,000. By contrast, their Native American neighbors experienced a near-annihilation caused initially by European diseases to which they had no resistance and by ecological destruction. The colonists were ruthless in exploiting rivalries between tribes and, regarding Native Americans as "savages," felt no qualms about encroaching on their lands and treating them brutally, while reacting with astonished indignation at any retaliation. The crushing of resistance by the Jamestown colony in Virginia had its New England counterparts in the Pequot War of 1637 and "King Philip's War" in 1675–1676.

The colonists also participated in the iniquities of African slavery. The earliest form of servitude in mainland British North America was the relatively mild one of indentured labor, whereby an immigrant contracted to serve a master for a number of years in return for his or her passage to the New World. The first party of African slaves—twenty of them—were brought to Jamestown by a Dutch ship in 1619. Demand increased, especially in the South, where a plantation economy (at first based on tobacco) began to develop at an early date. The black population of the colonies also rose rapidly, establishing a substantial, if as yet culturally "invisible," presence.

Regional differences remained much in evidence. New England's economy was based on farming, fishing, shipbuilding, and trade. Its Puritan tradition remained strong, though the links between church membership and the right to vote were broken in the 1690s, and the old certainties were beginning to fade despite the outbreak of hysteria manifested in the infamous Salem witch trials of 1692.

The Middle Colonies had a mixed character with certain distinctive elements: Dutch, Swedish, German, and Northern Irish settlers; riverine access to the interior; the fur trade; and the Quaker influence. Large areas of the South (notably excepting North Carolina) were already given over to a plantation economy and the gentrified society and easy-going Anglicanism to which it gave rise.

Despite the differences, the colonists were united in their attachment to self-government. There were assemblies in all the colonies, whether they originated as corporations or the possessions of royally appointed "proprietors." The only serious threat to their liberties, the autocratically run Dominion of New England set up by James II (the former duke of York), was immediately dismantled when the colonists learned of his dethronement in 1688.

Gradually the number of colonies grew to thirteen. New Hampshire received its charter in 1680. In 1702 East and West were united into a single New Jersey. North and South Carolina were separated in 1719. In 1733 the last of the thirteen, Georgia, was founded by General James Oglethorpe as a part-military and part-philanthropic venture.

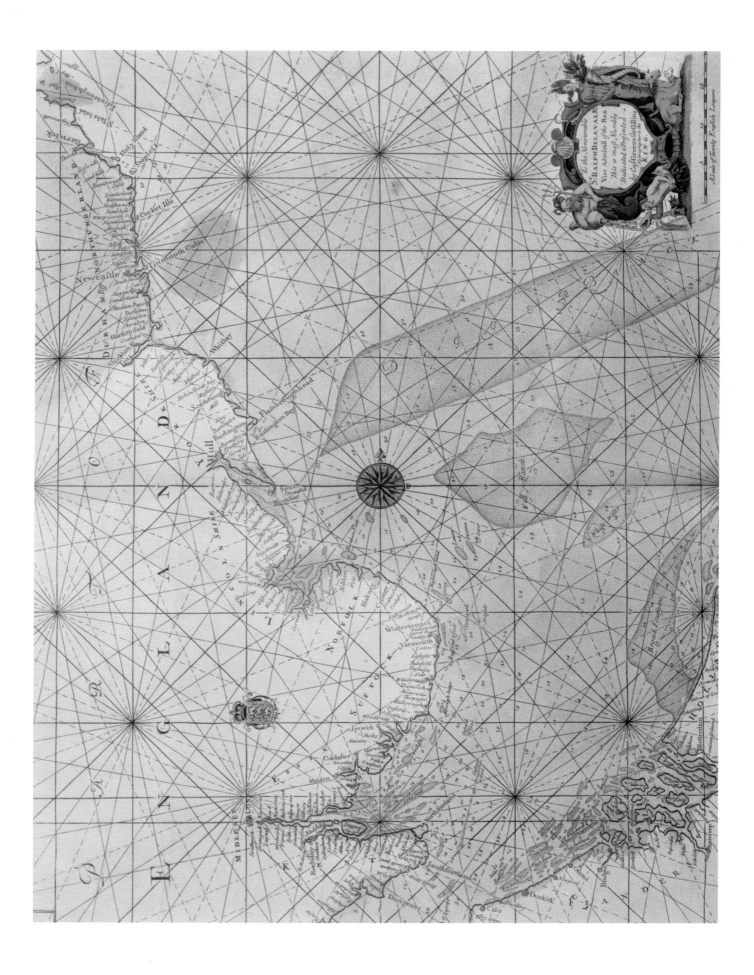

THE NORTH SEA
GREENVILE COLLINS

For a seafaring nation, the British were remarkably slow to produce charts for their own coastal waters, let alone for the seas and oceans beyond. Instead, they relied on the work done by their main naval and commercial rivals, the Dutch.

From the late sixteenth century, the printed charts, or "waggoners," used by British mariners were published translations of works by Lucas Waghenaer or his successors, notably the Blaeu and van der Keulen families. The first British collection of printed charts was John Thornton's *Atlas Maritimus*, which appeared in 1685, four years after the Admiralty at last realized the wisdom of sponsoring a survey of the British coasts. This was carried out by Captain Greenvile Collins, who labored over the following decade to produce the forty-eight charts published in

1693 as *Great Britain's Coasting Pilot.* Though not beyond criticism, Collins's charts were a significant advance. However, the promise that the Admiralty would take full responsibility for this vital activity was not fulfilled. Mariners in international waters continued to be dependent on Dutch charts, and at home Collins's work was not superseded for a century to come. As yet, it would have been impossible to predict the immense contribution to charting the oceans that would be made by British scientists, instrument makers, and explorers.

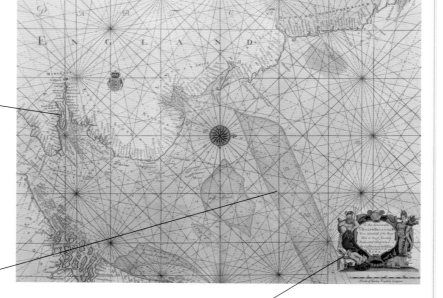

GATEWAY TO LONDON
Shoals and depth soundings are carefully marked on the Thames Estuary, the gateway to London. The city had already emerged as the great center of British maritime commerce.

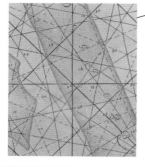

HARMLESS HAZARD
The Dogger Bank, a submerged sandbar in the middle of the North Sea, was a nautical hazard of the kind that accurate sea charts could render relatively harmless.

CARTOUCHE
The cartouche is executed with a pleasing naiveté, combining sea figures (including a self-consciously casual Neptune and trident) with images of rural plenty.

EFFECTIVE NAVIGATION

Coastal charts such as Greenvile Collins' made navigation safer and easier in European waters. But they were little help to mariners who undertook long voyages. Unable to determine their position accurately from day to day, they made landfall with difficulty and sometimes in the wrong place, until instruments became available for fixing latitude and longitude.

F or centuries, European ships plied mainly coastal waters, and captains or pilots survived rocks, shoals, and sandbars by using their knowledge of sailors' lore and local landmarks. There were also written sailing directions, descendants of the *periploi* used by the ancient Greeks, and, in the Mediterranean, the portolan charts that appeared later in the medieval period. In unknown waters there was

little to rely on except sharp eyesight and the hand lead that could be lowered to take depth soundings. Systematic charting began in the sixteenth century but was conducted (and revised) painfully slowly.

Hazardous ocean voyages

Voyages beyond the sight of land, though undertaken surprisingly often—by, among others, Vikings, Polynesians, and Arabs—were very dangerous. The great age of European exploration in the fifteenth century was made possible by the use of the magnetic compass and charts and the development of oceangoing ships. A limited mastery of the oceans was achieved as currents and wind systems became understood, but even roughly accurate location-finding was difficult on the open sea. Mariners were still reliant on dead reckoning—that is, fixing their position by estimating the direction and distance traveled from the last known position. The rate of sailing was assessed by experience, supplemented by the use of the log, a line with a piece of board

Above: The perils of seventeenth-century navigation: ships founder during a storm. Although mariners could fix their north–south position using heavenly bodies, their east–west location remained a mystery until Harrison devised his chronometer in 1735.

attached that, paid out behind the vessel, provided an indication of the relationship between time elapsed and distance traveled. The results would be noted in what became known as the ship's log. A number of distorting factors—for example, the curvature of the Earth—and, above all, the vagaries of the weather made dead reckoning a highly fallible method of navigation that could result in disastrous miscalculations.

The concept of determining geographical position by coordinates (latitude and longitude) had been formulated in antiquity by the Greeks Hipparchus and Ptolemy. The fifteenth-century rediscovery of Ptolemy revived it, but there were serious obstacles to its application at sea. Latitude—a ship's position north or south of the equator—could be fixed by observing the

angle of elevation above the horizon of the sun, the polestar, or other heavenly bodies. Longitude was more difficult to measure. Establishing a vessel's east–west position depended on using the precise time to read astronomical charts, but contemporary timepieces were unreliable on board a rocking ship.

Around the mid-fifteenth century, the Portuguese began fixing latitude at sea with the aid of a simple astronomical device, the astrolabe, which consisted of a metal disk with degrees marked on it. Over the next three centuries, instruments such as the cross-staff, backstaff, quadrant, and sextant steadily improved the accuracy of observations. Plotting a course became simpler with the wider use of Mercator's projection, and the charting of oceans and coastlines beyond Europe began.

Fixing longitude remains elusive

Determining longitude at sea remained almost impossible, however, since existing methods were hopelessly laborious or unreliable. On long voyages many vessels made landfall by sailing to the appropriate latitude and then maintaining a steady course until they arrived at their destination.

The main difficulty lay in correlating celestial observations with local time on Earth. Accurate timepieces were made from the second half of the seventeenth century, and they did make it possible to fix the location of terrestrial features. But on board ship, changes in temperature and gravity, and the motion of the vessel, conspired to disable every type of mechanism that was installed.

The matter was so important to the seafaring British that in 1714 an Act of Parliament offered a prize of £20,000 for the first workable method of determining a ship's longitude. The problems were solved over a number of years by the genius of a Yorkshire clockmaker, John Harrison. He completed the first of five superb chronometers in 1735, and after a forty-year struggle with an erratic officialdom, was finally awarded the prize in 1776. Scientific navigation was now a reality, and there would be no further substantial advances in this field until the twentieth-century introduction of radio- and radar-based technologies.

Above: The sextant was one of a number of instruments that improved navigation. This early eighteenth-century statuette represents a young British midshipman taking a sighting.

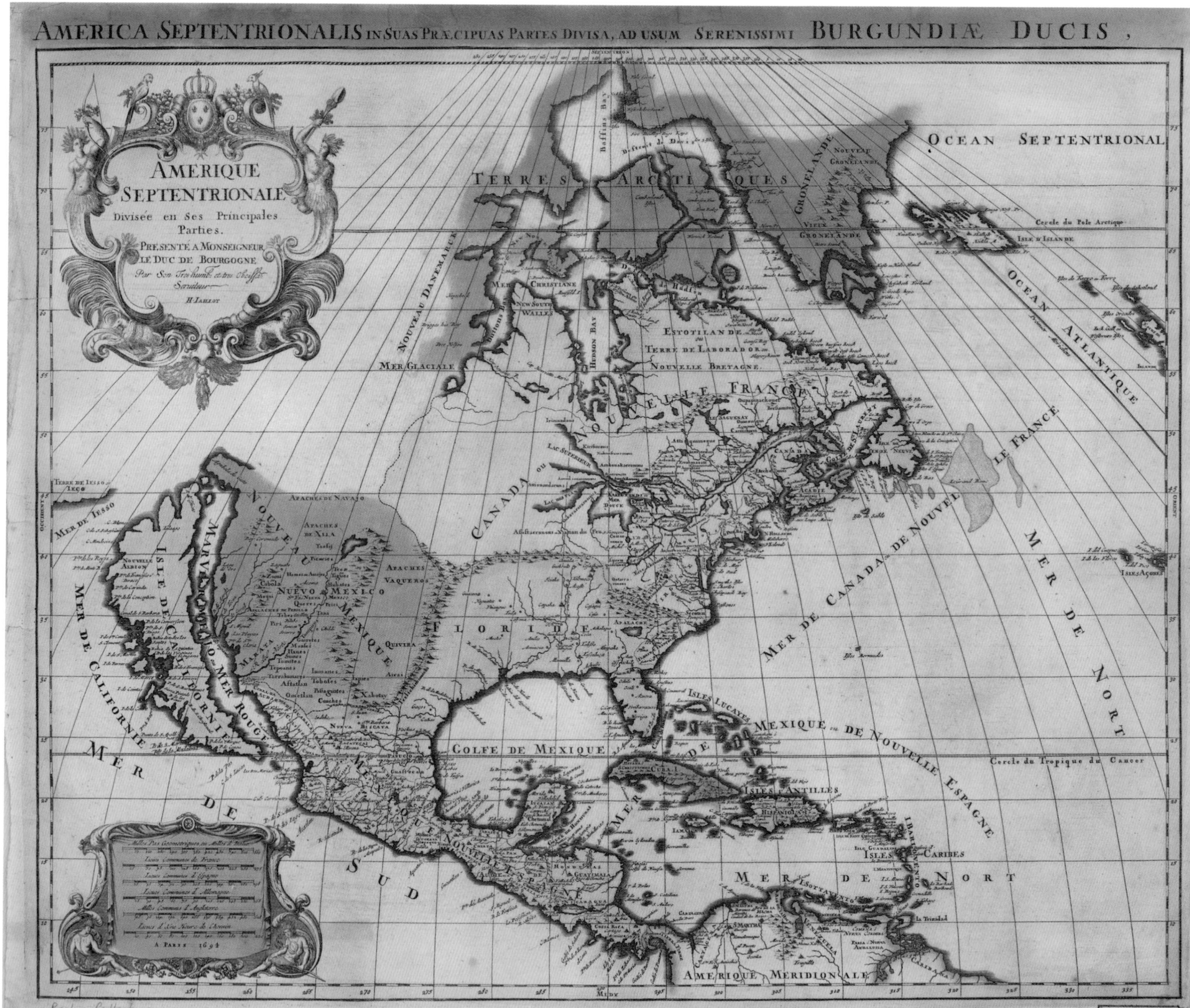

OCEAN SEPTENTRIONAL

OCEAN ATLANTIQUE

MER DE NORT

GROENLANDE

TERRES ARCTIQUES

NOUVELLE FRANCE

MER DE CANADA DE NOUVEL DE NOUVELLE ESPAGNE

Baffins Bay

Hudson Bay

MER GLACIALE

NOUVEAU DANEMARCK

CANADA ou

FLORIDE

GOLFE DE MEXIQUE

MER DE MEXIQUE ou DE NOUVELLE ESPAGNE

AMERIQUE MERIDIONALE

MER DE NORT

NOUVEAU MEXIQUE

MEXIQUE

NOUVELLE ESPAGNE

MER DE SUD

MAR VERMEIO ou MER ROUGE

ISLE DE CALIFORNIE

MER DE CALIFORNIE

AMERIQUE
SEPTENTRIONALE
Divisée en Ses Principales
Parties.
PRESENTÉ A MONSEIGNEUR
LE DUC DE BOURGOGNE
Par &c
H. Iaillot
A PARIS 1694

NORTH AMERICA
ALEXIS HUBERT JAILLOT

This French map of North America demonstrates the scale of national ambitions in the region, and also the problems of European cartographers, who were dependent on the reports of explorers and colonial officials. Jaillot's map nevertheless continued to be issued, with some corrections, until the end of the eighteenth century.

Jaillot was a sculptor who turned to cartography after his marriage to the daughter of a well-known map colorist. He went into partnership with the sons of Nicolas Sanson, the most distinguished French mapmaker of the previous generation, and prospered by issuing revised Sanson works as well as original maps. In 1672 the destruction by fire of the rival Blaeu publishing house gave him further commercial opportunities. Dated 1694 and dedicated to the Duke of Burgundy, Jaillot's map of North America claims vast areas for Canada, or New France. He appears to have been unaware of a great French achievement a dozen years earlier—La Salle's voyage down to the mouth of the Mississippi. On the map there is only a smaller river system, vaguely similar to the lower reaches of the Mississippi. In some other respects, too, Jaillot was behind the times, but the attractiveness of his map led to it being issued as late as 1792, still showing vanished New France and New Holland.

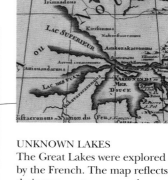

UNKNOWN LAKES
The Great Lakes were explored by the French. The map reflects their east–west progress: the limits of Lake Superior are unknown, and Michigan is represented by a misplaced Lac des Puans.

PASSAGE
The map leaves open the long-dreamed-of possibility of a passage into the Pacific, in this instance from Hudson's Bay into a large northwestern area, the precise nature of which—land or water—remains undefined.

IESSO
An imagined Terre de Iesso has replaced mythical Anian to create a strait into the Pacific. The other side of the strait is formed by the "island" of California.

DATED
New Sweden and New Holland (Pays Bas) are marked on the map, although the Dutch had annexed the Swedish settlement in 1655, only to give up all their territories to the British in 1674.

THE FRENCH IN NORTH AMERICA

Jaillot's map unwittingly shows North America just before one of its great defining features was discovered. La Salle's voyage down the Mississippi can be seen as the culmination of French exploration in North America, which put on the map the St. Lawrence River, the Great Lakes, and the main course of North America's mightiest river.

Norman and Breton vessels were fishing off the coast of Newfoundland by the early 1500s, and in 1524 King Francis I sent an expedition to the area. But the first significant French discoveries were made by Jacques Cartier, whose three voyages (1534–1542) opened up the Gulf of St. Lawrence and the St. Lawrence River. This would remain the point of departure for French explorers for well over a century, though initial attempts at colonization failed, and conflicts with Spain limited French effectiveness elsewhere in North America

Renewed efforts early in the seventeenth century brought considerable rewards, mainly thanks to Samuel de Champlain. Pushing down the St. Lawrence River, he founded a settlement with a future at Québec in 1608, traveled south down the Richelieu River to what is now Lake Champlain, and also reached the eastern end of the Great Lakes. In 1634, just a year before his death, Champlain sent Jean Nicollet farther into the Great Lakes. Nicollet became the first European to see Lake Michigan, and he heard reports of a great waterway that again raised hopes of a passage through the continent to the Pacific.

Exploration in New France
In spite of this there was little officially backed French exploration, though Jesuit missionaries pressed on (to found Montréal in 1642, among other things), and many individual traders penetrated far north in quest of furs. Then in 1672 an ardent expansionist, Count Louis de Frontenac, was appointed governor of New France (Canada). The following year he established a forward post, Fort Frontenac, on Lake Ontario, and sent out an expedition to find the great waterway. Led by Louis Joliet and Father Jacques Marquette, the party crossed the Fox–Wisconsin portage from Green Bay to the Mississippi. They sailed down the river, catching sight of its junction with the Missouri as they passed, until they reached the confluence of the Mississippi and the Arkansas, some 400 miles (650 km) from the Gulf of Mexico. At this point they turned back, partly because they were approaching Spanish

Above: The present-day Chicago skyline dominates Lake Michigan. The lake was first seen by Europeans when Jean Nicollet stumbled on it while searching for a waterway that would lead west to the Pacific Ocean.

territory and partly because they had realized that the south-flowing Mississippi was not the hoped-for gateway to the East.

Frontenac found a like-minded spirit in Robert de La Salle, who dreamed of creating a vast political and commercial empire in America. Given command of Fort Frontenac, La Salle was soon hatching new schemes, launching a cargo ship on Lake Erie and energetically building forts. He sailed the length of the Mississippi, reaching its mouth in April 1682, and claimed the entire river basin for France as "Louisiana." By this time Louis XIV was prepared for a colonial war, and La Salle was given command of an anti-Spanish expedition to seize the mouth of the

Mississippi. But he failed to find it, landing far to the west at Matagorda Bay, and after several unsuccessful forays, his remarkable career came to an end when he was murdered by his own men.

In the next few years, French settlements were established at Mobile, New Orleans, and elsewhere on the Gulf. As in Canada, this was to have more lasting cultural than political impact. French settlements also advanced from the north, reaching Ohio, but in the war of 1755–1763 France's possessions in mainland North America were lost to the British. Apart from a brief Napoleonic episode, France's imperial ambitions in the region came to an end, and with them its pioneering role in North American exploration.

Above: Bemused Native Americans look on as Robert de La Salle claims Louisiana—named for Louis XIV—for France in 1682. The region, comprising much of the vast Mississippi basin, was sold by Napoleon to the United States in 1803.

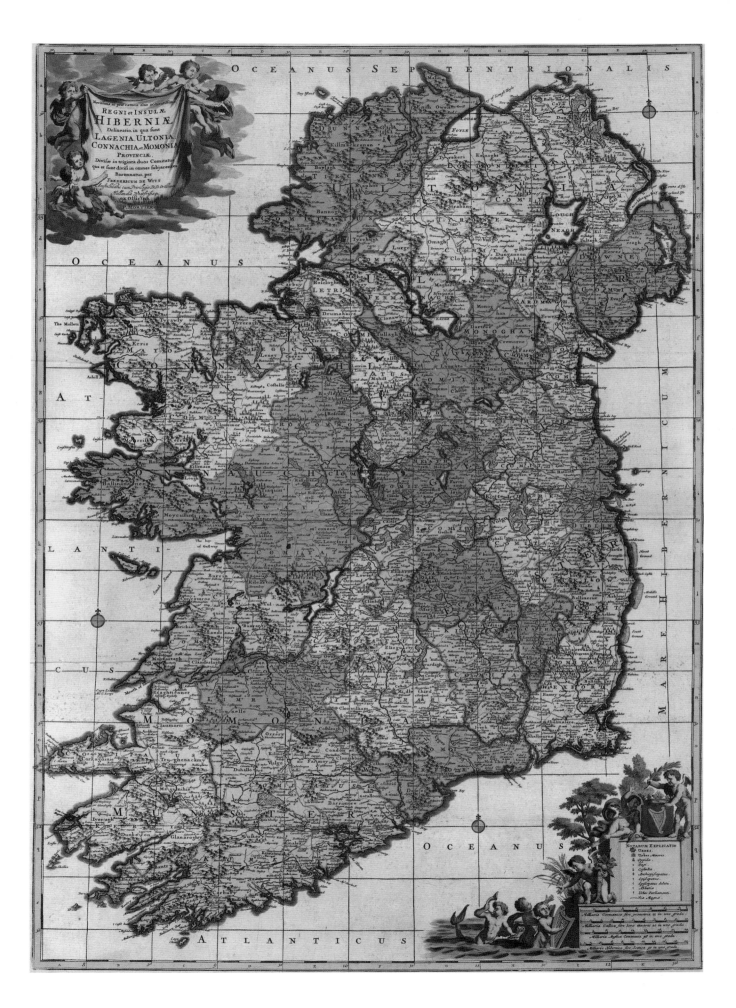

IRELAND
FREDERIK DE WIT

The great tradition of Dutch mapmaking was carried into the eighteenth century by the de Wits, father and son, who issued masterfully engraved and accurately detailed atlases, charts, and town plans. Based on a thorough English survey, Frederik de Wit senior's map of Ireland is information packed, as well as highly decorative.

Born in 1616, Frederik de Wit was apprenticed to Willem Blaeu before setting up his own Amsterdam printing house in 1648. He produced his first world map in 1660 for Hendrick Doncker's sea atlas, later superseded by de Wit's own atlas of charts, published about 1675 and famous in its day for its fine workmanship and accuracy. In the 1670s, when the Blaeu and Jansson firms went out of business, de Wit bought up many of their engraved copper plates and added them to his own stock.

After de Wit's death in 1689, his son, also called Frederik, became the head of the firm, in 1694 issuing an atlas of town plans based largely on Blaeu and Jansson publications. The younger Frederik died in 1706, when the stock was acquired by Pierre Mortier. Mortier has added his name to the title piece of the map of Ireland shown here, which was originally published about 1680. Mortier's descendants also continued to issue his maps, trading as Covens and Mortier for another century or more.

HIBERNIA
The title, "Delineation of Hibernia" (Ireland), is displayed in typical de Wit fashion, held up like a banner by distinctly un-Irish cherubs who are borne up by a cloud.

CONNEMARA
The coast of Galway is one of the most beautiful areas of Ireland's celebrated west coast. The Galway coast is, if anything, more spectacularly rugged and ragged than de Wit's outline suggests.

DUBLIN
Dublin and County Dublin stand on an open coastal plain that has tempted many invaders. Dublin itself was founded by Vikings from Scandinavia in 840 and was for centuries a bastion of English rule.

EXPLANATION
Beneath the harp, symbol of Ireland, de Wit has placed a table explaining the symbols he has chosen to use on this map, with its incredible wealth of detail superbly arranged and engraved.

THE MAPPING OF IRELAND

European knowledge of Irish geography was patchy before the seventeenth century. De Wit was the first continental mapmaker to publish a satisfactory map of the country, but he and his successors based their works on a mid-century survey that was conducted to strengthen England's dominance of the island.

During the Middle Ages the British Isles were on the edge of the known world. Even their rough outlines were either not known or were wilfully distorted in order to squeeze them inside the curve of a circular *mappamundi*. The relatively accurate (and also rectangular) "Anglo-Saxon" map (page 20) shows Ireland askew, lying east–west so that it is at right angles to its larger neighbor.

Fourteenth- and fifteenth-century portolan charts never showed the British Isles with the level of accuracy achieved in mapping the home waters of the Mediterranean. Although both main islands were rendered rather perfunctorily, a good many place names appeared on them; in the case of Ireland, this reflected a well-established trade relationship with Italy.

Early printed maps, including editions of Ptolemy's *Geography*, were not a significant advance on their manuscript predecessors when it came to representing Ireland. The outline of the country began to assume a form closer to reality in Gerardus Mercator's 1564 map of the British Isles, subsequently incorporated in the atlases published by Ortelius and Mercator himself. This was followed in the late 1580s by a particularly decorative map of Ireland alone, drawn by an Italian, Baptista Boazio, and included from 1602 in posthumous editions of Ortelius's atlas. Then maps by John Speed,

Above: Oliver Cromwell and his troops storm the Irish city of Drogheda in 1649. Many of the inhabitants were slaughtered, along with the troops defending it. Cromwell's merciless repression of the Catholics led to centuries of bitterness.

Above: The edge of Europe: the Atlantic Ocean seen from Clifden Bay on the rugged Galway coast. Ireland's remoteness from
the early modern centers of civilization meant that detailed knowledge of its geography eluded cartographers for centuries.

made for his *Theatrum*, drew on some early survey
work in Ireland and, characteristically, included a
variety of information along with images of national
types. As well as a general map of Ireland, Speed
included separate maps of the four historic provinces
(Ulster, Leinster, Connaught, and Munster) with inset
plans of the most important towns. Like other Speed
maps, these were copied by the Blaeus and other
publishers and so became known throughout Europe.

The rebellious north is tamed

By this time the English grip on Ireland, uncertain
throughout the Middle Ages, had perceptibly tightened,
and a major center of resistance had been eliminated
when once-wild Ulster had been "planted" by English
and Scottish Protestants. Nevertheless, there was a
Catholic rising in 1641 that helped to ignite the English
Civil War. When the war ended with the victory of
Parliament, an English army led by Oliver Cromwell
crushed the rebellious Catholics with a merciless
severity that was never forgotten. In the aftermath the
Catholic landowning class was largely destroyed and a
new Protestant Anglo-Irish establishment was created.

One consequence was a survey of the confiscated
lands in the 1650s, entrusted to the physician to the
parliamentary army, William Petty. A genuinely multi-
gifted individual, Petty persuaded the authorities to
extend the work to the entire island and carried it out
with great efficiency, employing a force of a thousand
assistants. Petty subsequently achieved distinction as
an inventor and, above all, as an economist; he was
the precursor of the "classical" school and the first to
make rigorous use of available statistics. Knighted by
Charles II, Sir William Petty utilized the results of his
survey in the thirty-six-map *Hiberniae Delineatio*, the
first atlas of Ireland, published in 1685.

Petty's maps replaced Speed's as the models for
discreet copying by British and European publishers,
although some original work was done in the
eighteenth century by Herman Moll, the Huguenot
mapmaker John Rocque (who actually lived in Dublin
for a time), and his brother-in-law, Bernard Scale. In
the nineteenth century, though private firms
continued to issue maps, they were overshadowed by
the definitive Ordnance Survey maps of Ireland,
issued from 1833.

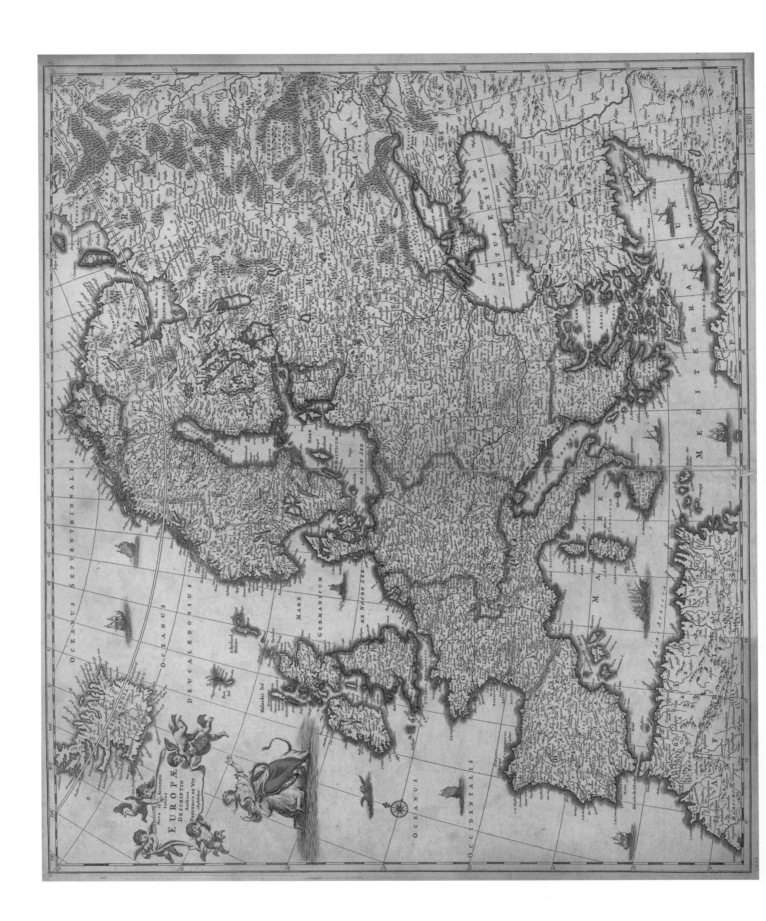

EUROPE
FREDERIK DE WIT

Before large-scale scientific surveys became feasible, cartographers had a difficult task evaluating limited and contradictory evidence. By the late seventeenth century, most of Europe had still not been rigorously measured and placed by coordinates. Much reliable information was available, however, and here the continent has assumed its familiar shape.

The map, which dates from around 1700, is not primarily a political one and only a large general view is given. Although Germany (the Holy Roman Empire) looks like a single state, in reality it was an agglomeration of territories, owing only nominal allegiance to the Holy Roman emperor; even the frontier lines are unrealistic, anachronistically including Switzerland and Burgundy. Italy, too, was divided into a number of states, some belonging to foreign powers. No frontiers at all separate the Low Countries.

England, Scotland, and Ireland are correctly shown as separate kingdoms, even though ruled by the same sovereign; formal unions would not take place until 1707 and 1800. Farther east, the Baltic is dominated by a Swedish empire taking in Finland, Ingria, and Livonia. Within a decade the rise of Russia and the building of St. Petersburg would transform the region. Finally, some rather arbitrary frontier-drawing conceals the vast area of the Ottoman Empire, its forces still not far from the gates of Vienna.

BOYS
This is one of a group of *putti*, naked babies or little boys, that carry the title cartouche. Sometimes winged, as here, they were common in classical art and inspired later styles.

ISLAND ERROR
Generally speaking, de Wit's map of Europe is very accurate. However, the island he places to the east of the White Sea (Album Mare) is in fact a peninsula of the mainland.

EUROPA
In Greek myth Europa, the daughter of the king of Tyre, was carried off and ravished by Zeus, king of the gods, who had disguised himself as a bull. Hence this and other versions symbolizing the continent.

PORTOLAN TRADITION
De Wit was a distinguished chartmaker, and although this map was obviously not intended for navigational purposes, his treatment of the coastline, with its overpronounced bays and inlets, is very much in the old portolan chart tradition.

PRINTED MAPS

The beauty of Frederick de Wit's maps exemplifies the skill of seventeenth-century Dutch engravers. The appearance of his works, and of other printed maps, was determined not only by the predominant style of the period but also by developments in the techniques employed to execute them.

Ancient maps were incised into clay tablets or drawn with inks or pigments on papyrus, parchment, or vellum (calfskin). Handmade and copied by hand (if copied at all), they were few in number and had only a very limited circulation.

From 1440 the invention of printing with movable type revolutionized map production and much else. Above all, it became economical to produce books in relatively large numbers using movable type (individually cut letters that could be assembled to print a page of type and then broken up and reused). Ideas and information, including maps, became available more rapidly and reached a wider audience. The first printed map was a basic T-O diagram, showing a tripartite world, in the 1472 edition of Isidore of Seville's *Etymologia*, an encyclopedic work dating back to the beginning of the seventh century that had enjoyed enormous prestige throughout the Middle Ages. It was followed in 1477 by the first printed edition of Ptolemy with maps. Both publications showed that printing was well suited to spreading errors as well as facts.

Printing by woodblock

These maps were printed by a method that was established in Germany even before the introduction of movable type. Pictures made by woodblock printing survive from the end of the fourteenth century, and the technique was probably used much earlier. (In the East it went back to the eighth century.) The craftsman began by drawing the design on a block of wood. Then the background—any area that was to remain white on the paper after printing—was cut away from the surface of the block, so that the design stood out in relief. When inked and pressed against paper, it printed the design, and the inking and printing could be repeated again and again to make multiple copies.

Woodblock printing was a convenient technique, since the block could be made the same thickness as

Above: The printing press developed by Johannes Gutenberg in the earlier fifteenth century allowed printers to combine text with woodblock printing to produce illustrated books.

the metal type. Both could be set within a single frame, making it possible to print the entire page (text plus picture or map) in a single operation.

Copperplate engraving

The characteristic products of woodblock printing were strikingly bold and "crude"-looking in a way that rather appeals to modern tastes. In the long run, however, the woodblock necessarily gave way to the copper plate, which made it possible to achieve much finer and more detailed effects. Whereas a relief technique was associated with woodblocks, copperplate engraving was based on the opposite, intaglio, principle. The design was cut into a sheet of copper, which was covered with ink. Then the surface was wiped over so that the ink remained only in the cut-away areas beneath it. When the plate was covered

with a sheet of paper and passed through rollers, the ink—that is, the design—was pressed on to the paper.

Copperplate engraving was particularly favored by Italian artists and craftsmen, and in the sixteenth century most of the leading mapmakers worked in Italian cities such as Rome, Florence, and Venice. In the seventeenth century the Dutch assumed the leadership, and increasingly refined craftsmanship and the rise of the spectacular Baroque style of art inspired some of the most beautiful maps ever made.

Copperplate continued to be the norm even after Dutch supremacy gave way to the less ornate art of British and French mapmakers. Significant technical change came only after 1800 with the introduction of steel engraving, mass-production lithography, and, ultimately, the scientifically precise techniques of the twentieth century.

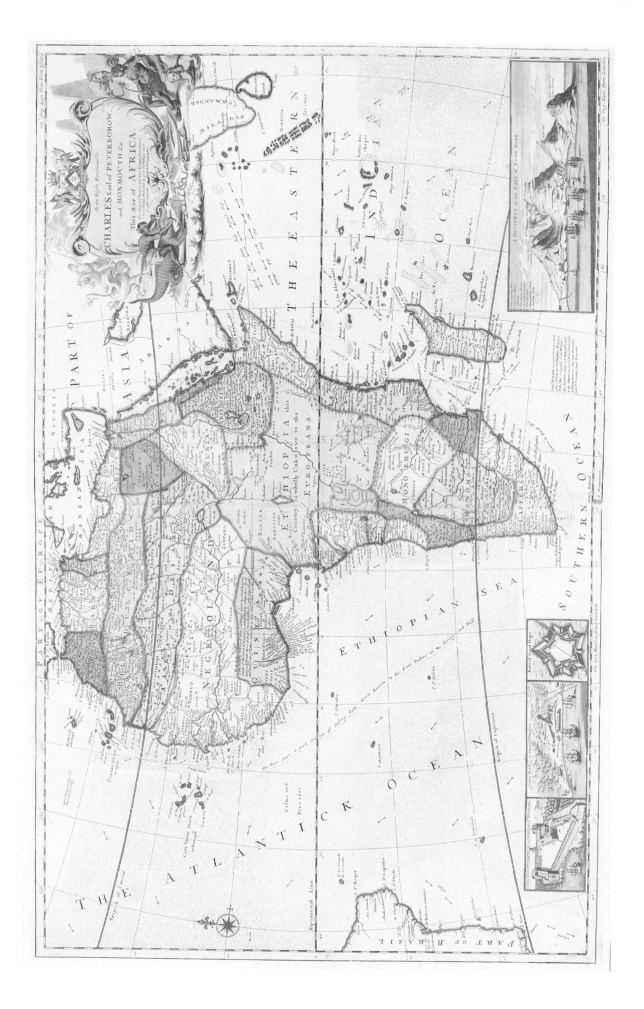

AFRICA
HERMAN MOLL

The Dutch engraver and cartographer Herman Moll settled in London, and his prolific output was primarily intended for a British audience. His change of residence was an almost symbolic act, taking place when Dutch naval power and the Dutch mapmaking tradition were waning and Britain's star was rising.

Moll's date of birth is not known. He is believed to have worked in England as an engraver from the 1680s before setting up as a cartographic publisher about 1700. He issued a great many maps, engraved by his own hand, including collections of British county and road maps, contemporary and historical atlases, and individual maps of different parts of the world. Many of these had marginal vignettes in the Dutch style, through which Moll is considered to have seeded the British cartographic tradition. He was one of the last cartographers to insist that California was an island. By the time Moll published his map of Africa in 1710, the outline of the continent was well known, but European cartographers relied on tradition and imagination for most interior details. Moll's map is in the more cautious eighteenth-century tradition, admitting to at least some ignorance and featuring in the vignettes the known coastal areas, of particular interest in view of Britain's growing imperial ambitions.

BERBER
"Barbaria" does not mean a barbarous land but the country of the Berbers. The Barbary Coast was a notorious haunt of pirates from the sixteenth century until the nineteenth-century conquest of Algeria by the French.

MEDIEVAL
In the beautifully drawn dedication, Africa is still represented by fanciful images reminiscent of medieval maps, notably a man subduing a crocodile, snakes in combat, and a native relaxing beside a very small lion.

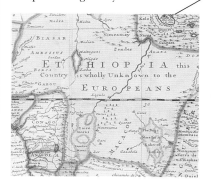

TERRA INCOGNITA
Moll's map distinguishes between Abyssinia and Ethiopia, which is used to describe Central Africa. Imagined lakes and rivers are included, despite the admission that "this Country is wholly Unknown to Europeans."

STRATEGIC
The "Prospect of the Cape of Good Hope" presents a detailed picture of the layout of a strategically vital Dutch settlement controlling the sea routes to India. Other vignettes emphasize the role of fortifications.

AFRICA: STILL IMAGINING

Moll's map of Africa was based on the best information available, but it also perpetuated some ancient errors. Like other cartographers, Moll was remarkably reluctant to abandon long-established ideas, though his map did at least admit that the heart of the continent was totally unknown to Europeans.

European ignorance of the interior is not really surprising. From ancient times most outsiders were deterred by the vast, dense, mysterious heart of sub-Saharan Africa, fortified by deserts, unhealthy, heat-saturated coasts, and mighty rivers with impassable rapids. Early investigators such as the historian Herodotus and the geographer Ptolemy had to base their statements on a handful of travelers' tales, with little to help them sift truth from fiction.

Understandably, the ancients were particularly fascinated by the location of the source of the Nile, a river that was immensely long and immensely important in making Egypt a land of abundance. Ptolemy concluded that it rose far to the south in the Mountains of the Moon and flowed into two lakes; out of these issued two rivers that united in Sudan before flowing north through Egypt to the sea.

Ptolemy was forgotten for centuries in the West, where *mappaemundi* invariably represented Africa—the Dark Continent—as the home of fabulous beasts and monstrous races. In a few instances use was made of superior Arab knowledge, and al-Idrisi introduced the Sicilian court to what was basically the Ptolemaic scheme, although with the addition of a second

Above: Around the African coast European slave traders built forts like this one, in Ghana. Here they accumulated captives delivered by slave traders from the interior—often rival African peoples—until there were enough to fill a ship across the Atlantic.

Above: Traditional Egyptian vessels called feluccas on the Nile. The source of the river, the basis of Egyptian civilization, held a particular fascination for European geographers.

"Nile," roughly corresponding to the Niger but sharing a common source with the north-flowing river.

Ptolemy's *Geography* acquired an immense authority after it was rediscovered during the fifteenth century. Even when Portuguese voyages round Africa disproved Ptolemy's belief that it was linked to the southern continent, representations of the interior continued to be based on his statements. As in mapping America, cartographers appear to have felt that anything passing for information was better than leaving a blank space. And where there was nothing to say, a range of mountains or a figure (Prester John being the favorite) served to amuse and distract the map-user.

A thirst for knowledge

During the sixteenth and seventeenth centuries, Europeans were eager for information about Africa, but the actual stock of knowledge hardly increased. Those who traveled to the continent were concerned with profit and survival—they stayed on the fort-protected coasts, to which African and Arab traders brought slaves, gold, and ivory from the interior.

Though the Portuguese were challenged by rival settlements of English, Dutch, German, and Danish traders, the situation remained essentially the same.

Nevertheless, maps of Africa multiplied and, thanks to the new invention of printing, reached a much wider readership. The first large-scale rendering of the continent was an eight-sheet map by Jacopo Gastaldi, published in 1564. It contained some new information (notably the existence of the Zambezi and Limpopo rivers and the great walled city of Zimbabwe), but spaces were filled by multiple chains of mountains, and the Ptolemaic scheme was retained, with the legendary source lakes now named as Zaire and Zaflan.

Among the few genuine explorers were Jesuit missionaries to what was then known as Abyssinia. (Confusingly, the country's present name, Ethiopia, was used to describe a much wider and vaguer region of the interior.) Their reports inspired the Abyssinian studies of Hiob Ludolf, whose map of 1683 correctly identified Lake Tana as the source of the Blue Nile. Such was the strength of tradition that Ludolf's findings were scoffed at for a generation or more.

The fading of fantasy

More rational views began to prevail in 1700, when Guillaume Delisle published his map of Africa. In the tradition of French mapmaking that would dominate the eighteenth century, Delisle dispensed with monsters and left blanks where information was lacking. The two lakes disappeared and he expressed his doubts about the Nile–Niger link. Herman Moll seems to have been aware of continental trends, and modified his approach accordingly. Nevertheless, as if to placate Ptolemy's ghost, he retained Zaire and Zaflan—reduced to the status of morasses—in the far south, with the Zaire fed by a kind of second or substitute Nile flowing right across Central Africa (called "Ethiopia" on the map).

Though eighteenth-century maps treated their subject with greater sobriety, modern exploration began only with the founding of the African Association in 1788 and the sponsored exploration of the Niger by Mungo Park. These events initiated a century of epic journeys that would at last make it possible to map, instead of imagine, the Dark Continent.

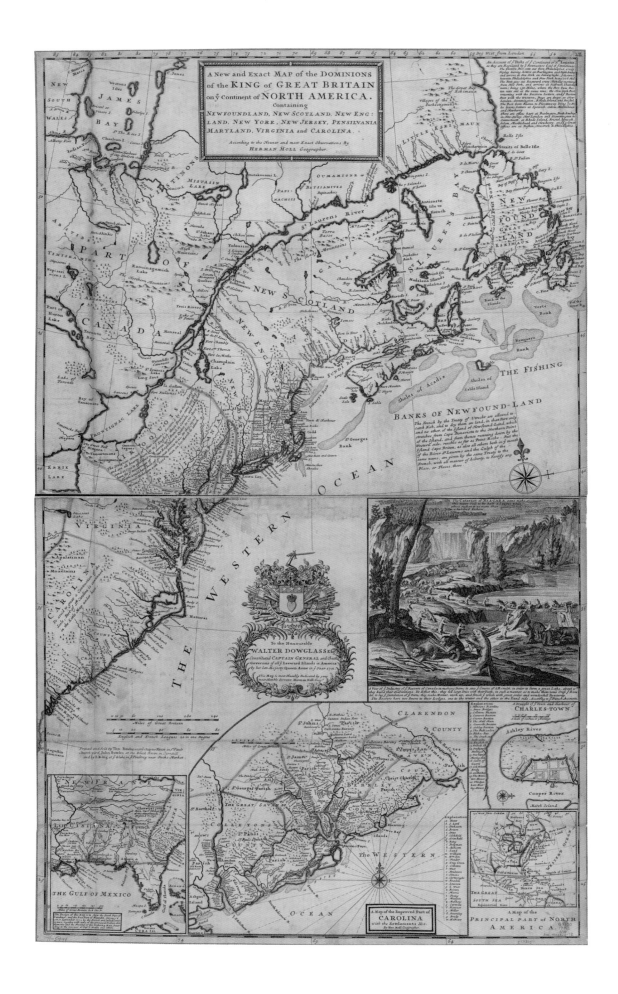

THE "BEAVER" MAP
HERMAN MOLL

The map acquired its nickname from the large inset with amusingly anthropomorphized beavers. Despite this fanciful touch, Moll's work is well designed and engraved, and packed with useful information that no doubt contributed, as much as the beavers, to its popularity. Published in 1715, it shows the colonial situation following the 1713 Peace of Utrecht.

The Peace of Utrecht ended the War of the Spanish Succession, which for most of the combatants was primarily a European conflict. In America, however (where it was known as Queen Anne's War), the war marked a serious French reverse: Acadia was lost, becoming New Scotland (finally Nova Scotia), and English possession of Hudson Bay and Newfoundland was recognized. Lake Frontignac had not yet become Lake Ontario, however, and the French presence to the north and west remained formidable. Published as two sheets, the main map was framed in such a way as to omit most of New France and include a large expanse of the "Western Ocean," which was filled with the dedication, a vignette, maps that clarified the great-power situation, and a plan of "Charles Town" (Charleston). The delightful large vignette shows model colonists—industrious beavers forming a production line within sight of "the cataract of Niagara" in order to build a dam.

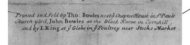

DEDICATION AND TROPHY
Moll's so-humble dedication to the governor of the Leeward Islands is typical of the relations between creators and actual or prospective patrons. The heraldic trophy is splendidly executed.

WONDER WORKERS
This inset of busy beavers gave the map its nickname. Remarkable though these creatures are for their dam-building, they are not known to shoulder bundles of sticks or transport objects on their tails.

ON SALE
The homely inscription here is not unlike those on earlier English maps (for example, Speed's). Map- and book-publishing remained trades of restricted scope, closely associated with the firms that printed them.

COLONIAL EMPIRES
This neat little inset sums up the colonial situation in 1715, with Britain dominating the eastern seaboard, France claiming vast territories in Canada and the unsettled continental center, and Spain holding on to the rest.

STRUGGLE FOR MASTERY

Moll's map of Britain's North American
colonies contains many references to
Anglo-French rivalries, still unresolved at
the time of publication in 1715. At first a
sideshow, the colonial wars became
increasingly important both to the home
governments and the settlers, ending with
what appeared to be a clear decision
in Britain's favor.

For much of the seventeenth century, France and
England were at peace in Europe, and contacts
between their American colonies were too slight
to breed hostility. However, in time French claims to
the Mississippi Valley and adjacent territories would
probably have led to conflicts with the colonists on the
eastern seaboard. The French presence in "Louisiana,"
primarily military and commercial, denied the rapidly
multiplying British colonists access to fertile, thinly
populated lands and everything that lay beyond them.
In Europe, too, French ambitions were likely to
provoke countermeasures by the rising commercial
and maritime power of Britain.

War in Europe and the colonies

The beginning of the struggle was brought forward by
Britain's "Glorious Revolution" of 1688, which
overthrew James II and put William and Mary on the
throne. As leader of the Dutch, William was committed
to opposing France, and he quickly brought Britain
into a grand coalition against Louis XIV. The hard-
fought War of the League of Augsburg (1689–97) had
its colonial counterpart, known as "King William's War,"
and both ended without significant territorial changes.

The New England colonists' experience of the
destruction of their shipping, and bloody raids and
massacres by the French and their Native American
allies, ensured their wholehearted participation in
such conflicts. During Queen Anne's War
(1702–1713)—the colonial phase of the War of the
Spanish Succession—a colonial force took part in the
capture of Port Royal (now Annapolis Royal), the
principal town in Acadia. The terms of the peace
treaty represented a blow to French hopes: Acadia had

to be given up, and British claims to Newfoundland
and Hudson Bay were conceded.

New conflicts proved indecisive. Spain, though no
longer a first-class power, proved able to defend its
Floridan possessions when war with Britain broke out
in 1739. As the Anglo-Spanish conflict merged into a
wider European war (the War of the Austrian
Succession), Britain and France again came to blows
in the Old and New worlds. The most spectacular feat
of "King George's War" was the capture in 1745 of a
French superfortress, Louisbourg, on Cape Breton
Island, by a mixed colonial and regular force backed
by the British navy. Other campaigns were less
successful, and when peace was made in 1748, the
status quo in both Europe and America was restored.

The French go on the offensive

In the course of the wars, New England and New York
had suffered cruelly. From the colonial point of view,
the situation was becoming critical as the prospect
beckoned of expanding across the Appalachian
Mountains. In 1750, when the newly formed Ohio
Company began exploring the upper Ohio River
region, the French responded with a fort-building
program. In 1753 the young George Washington was
sent to demand a French withdrawal. As it was not

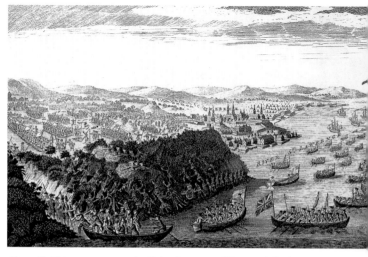

Above: British troops cross the Saint Lawrence River and deploy
on the Plains of Abraham. The subsequent battle for Québec in
September 1759 determined the destiny of Canada.

Above: Native American allies of the French ambush Britain's general Edward Braddock near Fort Duquesne. The promise of lucrative trade convinced a number of peoples of the northeastern woodlands to support the French rather than the British.

forthcoming, he returned in the spring of 1754 and constructed Fort Necessity as a counter to the French Fort Duquesne. But when the French attacked in force, Washington was compelled to surrender.

This effectively began the French and Indian War, which was fought for two years without a formal declaration. The British commander, General Braddock, was defeated and killed near Fort Duquesne, but no decisive result was achieved. More forts had been built than captured by June 1756, when hostilities between Britain and France became official, initiating the Seven Years' War in Europe. Britain made an unpromising beginning, and in America the French, led by the Marquis de Montcalm, captured a series of forts. Then in 1757 William Pitt the Elder infused the British war effort with a new energy and pursued a strategy in which colonial operations assumed a much higher priority. By 1758 the British were advancing in America, and 1759 was the "Wonderful Year" of victories on all fronts. In America the knockout blow was the famous battle on the Plains of Abraham at which both Montcalm and Britain's General Wolfe lost their lives and Québec fell.

When peace was made in 1763, France surrendered all of Canada and Louisiana east of the Mississippi to Britain; Spain ceded Florida to Britain and took over the French possessions west of the Mississippi. Since Spain was far less formidable than France, British North America was secure at last—perhaps, as some anxious British pamphleteers suggested, too secure to continue for long in a dependent situation.

Above: Death of a British hero: The victorious general James Wolfe, architect of the capture of Québec, dies in battle on the Plains of Abraham beside the city.

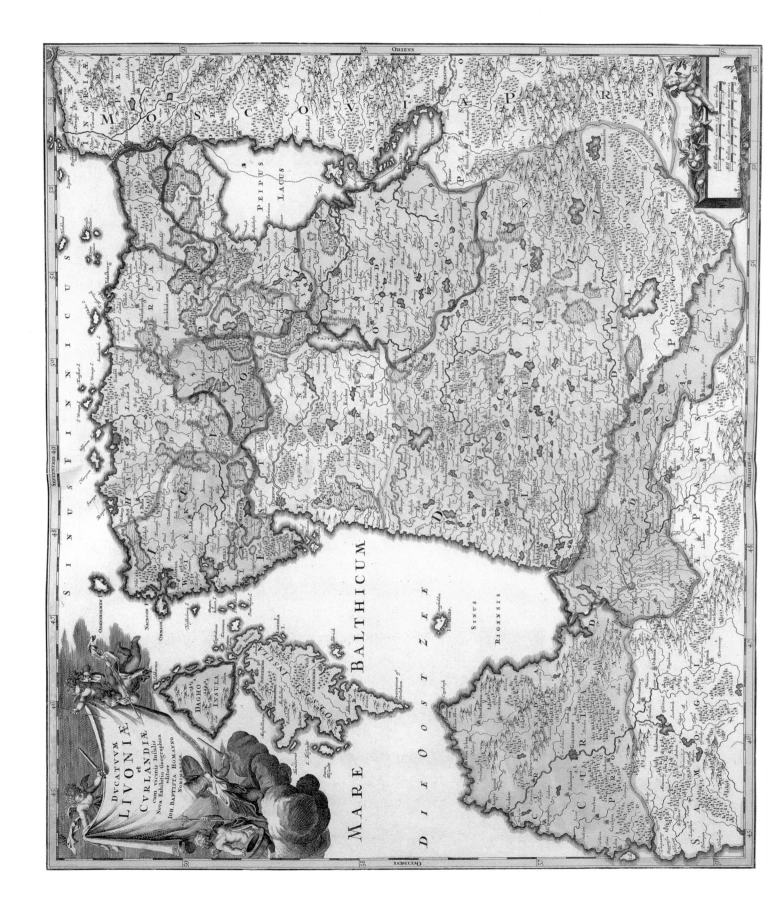

MOSCOVIÆ PARS

INGRIÆ PARS

SINUS FINNICUS

MARE BALTICUM

DIE OOST ZEE

SINUS RIGINSIS

LIVONIA

PEIPUS LACUS

CURLANDIÆ PARS

SEMIGALLIÆ PARS

DAGHO INSULA

DUCATUM LIVONIÆ
et
CURLANDIÆ
cum vicinis insulis
Nova Exhibitio Geographica
editore
IOH. BAPTISTA HOMANNO
NORIB.

LIVONIA AND COURLAND
JOHANN BAPTIST HOMANN

This map is the work of Johann Baptist Homann, the leading eighteenth-century map publisher in Germany. It gives a political and physical account of two much fought-over Baltic duchies that had already changed hands several times during great power conflicts in northeastern Europe.

Homann worked as an engraver for map publishers in Amsterdam before starting his own business at Nuremberg in 1702. By the time of his death in 1723 he had published five atlases and many single maps. His son Johann Christoph carried on until his death in 1730. He left the business to his brother-in-law on condition that he and his descendants traded as "Homann's Heirs." Under this name the firm survived until 1813, its production culminating in an *Atlas Maior* of 1780 with 300 maps.

The example of Homann's work shown here was published in his *Neuer Atlas* (c. 1714). The duchies of Courland and Livonia (roughly modern Latvia and Estonia) had a turbulent history, most recently in long struggles for control of the Baltic between Poland, Sweden, and Russia. When the map was published, Russia had emerged as a great power, defeating the Swedes and seizing Livonia. But it was not then clear that the Russian victory was a decisive one, and the map is discreetly silent on matters of sovereignty.

BABES IN ARMS
The titles of Homann maps were often accompanied by figures representing the occupations of the inhabitants. But here, perhaps prompted by the many wars fought over Livonia and Courland, the figures are armed and aggressive cherubs.

PRESENT-DAY FRONTIER
The Russo-Livonian frontier is formed by Lake Peipus, Lake Pskov to the south, and, to the north, the River Narva flowing into the Gulf of Finland. An almost identical frontier now separates Estonia from Russia.

LANDSCAPE
Woods, mountains, marshes, and lakes are all represented in this section of the map, not by contours or symbols but in a less precise but decorative pictorial mode.

CAPITAL
Riga was a thriving city by the thirteenth century, when it became part of the Hanseatic League that dominated the Baltic. Later annexed to Poland, Sweden, and Russia, it is now the Latvian capital.

SWEDEN: RISE AND FALL

In the seventeenth century contests for supremacy in the Baltic reached their climax in a series of great wars. Swedish feats of arms dazzled Europe and brought fame to a line of warrior-kings. However, Sweden's Baltic empire proved to be a fragile structure that hardly survived the century.

The Baltic region was important to the European economy, exporting grain, timber, and naval supplies, and cities such as Riga were lucrative prizes for empire-builders in the region. Early in the seventeenth century the most favored of these seemed to be Denmark, which controled the Sound—the narrow entrance to the Baltic—and what is now southern Sweden. Poland was larger and potentially strong but handicapped by the weakness of its monarchy. Brandenburg, though poor in resources, acquired Prussia by marriage in 1618 and nurtured regional ambitions. Russia's briefly held Baltic territories had been lost, and the country's technological backwardness and internal "Time of Troubles" appeared to put it out of contention.

A great power emerges

Sweden was not particularly favored by geography, and the country had only a small population, few resources, and, with the exception of Finland and part of Livonia, few possessions. In 1611 the sixteen-year-old Gustavus Adolphus became king. His position was so weak that almost at once he was forced to make large concessions to the nobility and to end a war with Denmark by paying an indemnity. Yet within a few years Gustavus's organizational ability and military genius had turned Sweden into a great power. An efficient administration was set up, the army and navy were remodeled, and, as Sweden's rich mineral wealth was exploited, an armaments industry was developed to keep the king's armies supplied during his European adventures.

These began with a war against embattled Russia in 1617, which enabled Sweden to consolidate its holdings on the Baltic and again deprive the Russians

of access to the sea. Then four years later Gustavus besieged and captured Riga, initiating an eight-year conflict with Poland from which Sweden emerged with all of Livonia.

These were arguably Gustavus's most solid achievements. However, his European renown was the result of his decision to intervene in Germany, the main battlefield of the Thirty Years' War (1618–1648). In 1631 the Holy Roman emperor and the Catholic cause seemed to have triumphed in their struggle against their Protestant opponents, initially led by Bohemia and the Palatinate. Sweden's Danish rival

Below: Gustavus Adolphus was a military genius who became the champion of Protestantism in the Thirty Years' War.

had attempted to intervene on the Protestant side and had been crushed. Now Gustavus, having decided that he would become the Protestant champion, launched a new phase of the war when he landed on the Pomeranian coast and conducted a whirlwind campaign that carried his forces to the Rhine. In 1631–1632 his two great victories at Breitenfelde and Lützen transformed the European situation. Although Gustavus himself was killed at Lützen, the Swedes continued to take part in the war, and when peace was made in 1648, they registered further gains on the Baltic and in Germany.

The now impressive Swedish empire was maintained and even expanded over the fifty years following Gustavus's death. In 1660 Denmark ceded its remaining possessions in the south of the Scandinavian peninsula. This region—now thought of as an integral part of Sweden because of its location on the

Above: The riches of the Baltic—fur, timber, and fish—were the basis of a wealthy regional economy.

Above: After his defeat by the Russians at Poltava, Charles XII of Sweden fled to the Ottoman Empire, where he remained, intriguing ineffectively, for five years.

peninsula—proved to be the most permanent of its conquests. The fragility of the other Baltic possessions was demonstrated in the 1670s, when Brandenburg overran Pomerania. It was restored, but only on the insistence of Sweden's French allies.

In retrospect the loss of Sweden's coastal "empire" seems inevitable, yet the direct cause was the almost lunatic recklessness of another royal military genius, Charles XII, who became king of Sweden in 1697. Early in his reign only the genius was on display. In 1700 Charles, faced with a hostile Danish–Polish–Russian coalition, thrust swiftly into Zeeland and knocked Denmark out of the war. In November 1700 he attacked Russia and decisively defeated the czar, Peter the Great, at Narva.

The empire collapses

But then Charles spent six years campaigning in Poland, while Peter rebuilt his army, advanced into the Baltic, and founded a new capital, St. Petersburg, at the mouth of the Neva. When Charles finally marched into Russia, he found himself struggling against the winter and vast distances, Russia's traditional allies in the face of invasion. Finally he turned south into the Ukraine, and in 1709 besieged Poltava until Peter's relieving army arrived and inflicted a smashing defeat on the exhausted Swedes.

Charles fled south, taking refuge in Turkey. Now behaving erratically, he did not return to Sweden until 1714. In 1718 he was killed by a sniper's bullet during a campaign in Norway. Sweden's Baltic empire had vanished, and for two hundred years the region would be dominated by Russia and a newly emerging great power, Brandenburg–Prussia.

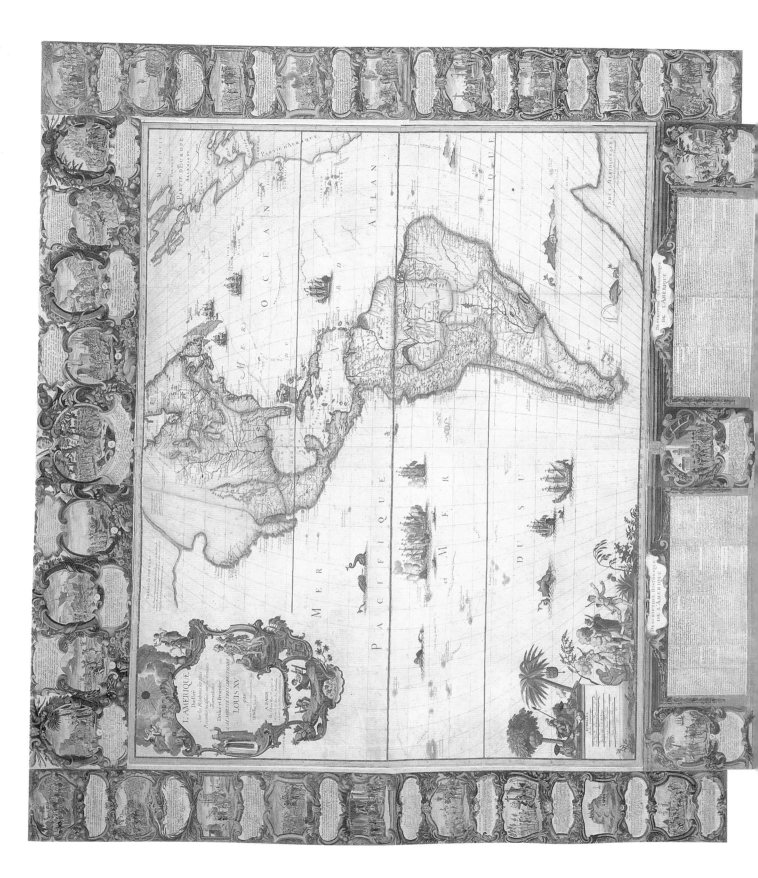

AMERICA
JEAN BAPTISTE NOLIN

Jean Baptiste Nolin was one of the many outstanding figures whose works made France supreme in eighteenth-century cartography. His very large and detailed wall map of the Americas was both a spectacular decorative object and a record of as yet unresolved imperial rivalries in the New World.

Measuring 54 by 48 in. (137 by 121 cm) and dedicated to King Louis XV, the map was produced in 1740 by Nolin the Younger. In 1708 he had inherited a flourishing map-publishing business from his father, J. B. Nolin the Elder, who seems to have thrived by questionable means, including copying others' works. The younger Nolin had a less controversial career, producing maps of the continents and preparing an *Atlas Général* that was not published until twenty years after his death in 1762.

His New World map was drawn at a time when France still entertained ambitions in North America, despite the loss of Nova Scotia to the British. Largely thanks to the endeavors of La Salle (page 191), France dominated the Mississippi basin, known as Louisiana, and had claimed or settled territories stretching from the mouth of the Mississippi to the Gulf of St. Lawrence. New editions of the map continued to be published despite the loss of Canada by France and the revolt of the Thirteen Colonies against Britain.

DEDICATION
The superb cartouche is dedicated to His Christian Majesty, Louis XV of France. It emphasizes the evangelization of the New World; elsewhere, scenes of violence tell a different story.

RIVALS
These vignettes picture (above) the English in Virginia and (below) the French in New France, representing violent and peaceful episodes of a type familiar in colonial history.

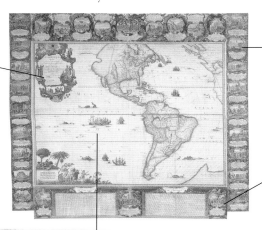

AT WAR
A deadly naval engagement, taking place among the familiar depictions of sportive sea creatures. It is one of several features suggesting the eighteenth-century intensification of colonial rivalries.

FANTASY
The picture shows Patagonian giants. Above it, on the map, the discovery of the "Southern Indies" is attributed to a Frenchman, Paulmier, who visited an unidentified land (probably Brazil) in 1503–1505.

THE SUPREMACY OF FRANCE

Taking over the leadership from the Dutch, eighteenth-century French cartographers pioneered a more scientific approach to their craft. They also employed leading artists at a time when the French decorative tradition was at its height, so that many of their maps have an enduring appeal.

French cartography was not greatly distinguished before the seventeenth century. French mariners were dependent on Portuguese charts until the 1540s, when a school of manuscript mapmakers developed at Dieppe, France. Among the school's productions were the Vallard Chart (page 76), showing French discoveries in Canada. Nicolas de Nicolay drew the first chart printed in France, and in 1594 regional maps by various hands were collected and published as the first national atlas at Tours by Maurice de Bouguereau.

The rise of a more scientific attitude toward mapmaking took place in conjunction with the consolidation of the French state in the seventeenth century by Cardinal Richelieu and King Louis XIV. The fount of patronage, a centralized and bureaucratic monarchy needed reliable information in order to control, mobilize, and tax its subjects.

The first royal cartographer, Nicolas Sanson, published well-researched maps ranging from the French waterways to *Le Canada ou Nouvelle France* (1656), which was the first to show all of the Great Lakes. Mapmaking was still very much a family business, and after his death in 1667, Sanson was succeeded by his son and grandson. Afterward the Robert de Vaugondys family took over the Sanson materials, using them in revised form for the *Atlas Universal* of 1757, sponsored by the king's mistress, Madame de Pompadour, and other members of the court.

Even more distinguished were the geographer and historian Claude Delisle and his sons, Guillaume, Simon, Joseph Nicolas, and Louis. The most gifted, Guillaume, was a prodigy who is said to have drawn his first map in 1684, when he was nine years old. In 1718

he became geographer to the king, and his scrupulously researched maps and atlases were used long after his death. His brothers Joseph Nicolas and Louis entered the service of Peter the Great, czar of Russia, as astronomers and surveyors. In 1745 Joseph Nicolas issued an *Atlas Russicus*, the first collection of maps devoted to a country that had only recently entered the European arena.

A host of gifted mapmakers

In the course of the eighteenth century, France produced many fine mapmakers. One of the most celebrated was Jean Baptiste Bourguignon d'Anville, who was only twenty-two when he published his map of France in 1719. During his sixty-year career, he remained in Paris, producing the most accurate maps

Below: When Cardinal Richlieu strengthened and centralized the French state, accurate maps became a national priority. How could bureaucrats function without adequate documentation?

of his time simply by diligent research and a critical approach to his sources. His *Nouvel Atlas de Chine* (1737), based on Jesuit and other European reports and surveys, was particularly outstanding.

Despite the general trend toward scientific sobriety, some French cartographers were apt to provide speculative information where nothing certain was known. Among the offenders was Guillaume Delisle's son-in-law, Philippe Buache, whose imaginative excursions included a map of the ocean beds, their topography deduced from the existence of islands as the mountaintops of the subaqueous world. A good

many eighteenth-century French maps (their numbers swollen by copying) show Australia extending far to the east to join New Zealand, central Canada terminating at the shores of a western sea, or eastern Siberia and Alaska subsumed in a vast subarctic landmass. All of these flights of fancy were discredited before the end of the century by the voyages of English explorer Captain James Cook.

Though many French cartographers created informative and beautiful maps, the most significant advances from a purely scientific point of view were made by a single family, the Cassinis (page 216).

Above: When François Boucher painted Madame de Pompadour, mistress of Louis XV, with an open book on her lap, he was making a shrewd point. She was intelligent and cultured, a real power in the kingdom and an important patron of artists—and mapmakers.

MAP OF FRANCE
CESAR CASSINI DE THURY

This soberly accurate map of France was one result of surveys carried out by
a single family of scientists. Four generations of Cassinis labored to conduct the first
detailed and accurate survey of an entire country. The 1745 map fixed the contours
and principal features of France, laying the foundations for an
even more comprehensive Cassini survey.

Although basic surveying techniques were known from ancient times, improved instruments and considerable political will were required to apply them on a national scale. This was first achieved by the centralized French monarchy of Louis XIV, spurred on by Giovanni Cassini and largely carried out by his descendants. The basic method used to measure distances was triangulation, which employed a combination of geometry and

trigonometry, described on page 237. A pioneering figure, Abbé Jean Picard, measured the arc of the meridian running though Paris, and then between 1679 and 1681 triangulated the French coastline. The result was that France was shown to be smaller than previously believed: Louis XIV lamented that the survey had cost him more territory than a war. These achievements led on to the Cassinis' even larger-scale project, only fully brought to fruition in 1789.

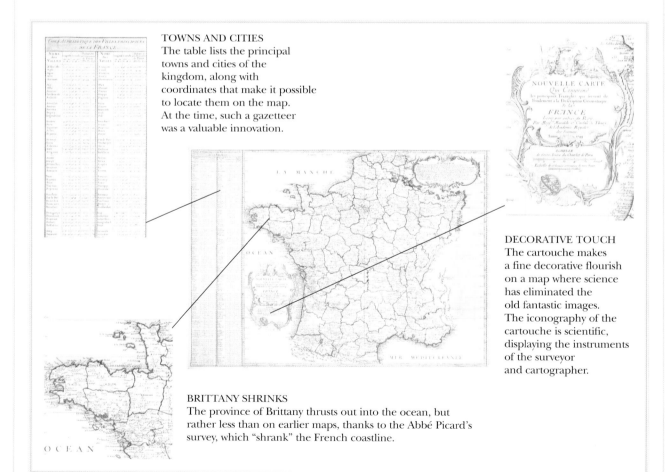

TOWNS AND CITIES
The table lists the principal
towns and cities of the
kingdom, along with
coordinates that make it possible
to locate them on the map.
At the time, such a gazetteer
was a valuable innovation.

DECORATIVE TOUCH
The cartouche makes
a fine decorative flourish
on a map where science
has eliminated the
old fantastic images.
The iconography of the
cartouche is scientific,
displaying the instruments
of the surveyor
and cartographer.

BRITTANY SHRINKS
The province of Brittany thrusts out into the ocean, but
rather less than on earlier maps, thanks to the Abbé Picard's
survey, which "shrank" the French coastline.

THE CASSINI FAMILY

For more than a century, the Cassinis were at the heart of scientific developments in France. Nor were their achievements restricted to a single country: They initiated cooperation between traditionally hostile France and Britain, and in time every state followed their example in undertaking a comprehensive national survey.

Giovanni Domenico Cassini was an extraordinarily gifted young man who became professor of astronomy at Bologna University in Italy in 1650, when he was only twenty-five. He made a number of significant discoveries in astronomy and mathematics, and in 1669 Louis XIV of France

Above: The scientist Giovanni Domenico Cassini, celebrated in this engraving under the French form of his name. Among other things, he discovered four moons of Saturn.

invited him to become director of the newly established Paris Observatory. He subsequently took French citizenship, and became known as Jean Dominique Cassini.

Even before leaving Bologna, Cassini had made an important contribution to cartography. In his *Ephemerides* he tabulated the movements of the moons of Jupiter, first observed by Galileo Galilei. In conjunction with an accurate timepiece, the tables made it possible to determine the longitude of an observer's position on land. (Determining longitude at sea remained a problem for another century.) This great advance proved its significance when Cassini began work on his famous world map, for which correspondents all over the world sent him observations based on his own tables. Consequently, his *Planisphère Terrestre*, published in 1696, was the most scientifically accurate world map of its time and a model for future endeavors.

Meanwhile Cassini was involved in Abbé Picard's triangulation surveys and advocated extending them to all of France. After Picard's death, Cassini was appointed to carry on the work, but fluctuating royal enthusiasm caused long delays. Though Jean Dominique did not die until 1712, it was left to his son Jacques to take over the near-dormant project as well as succeeding his father as director of the observatory.

Cassini versus Newton

As director, Cassini became drawn into an Anglo-French scientific dispute. The English scientist Sir Isaac Newton had argued that a spinning globe must bulge slightly at the equator and be slightly flattened at the poles. If he was correct, the consequence for cartography would be that the farther north a measurement was taken, the longer a meridional degree would become. Jacques Cassini was convinced that the reverse was true, and encouraged Louis XV to fund expeditions to the Arctic (Lapland) and the equator (Peru) to make the appropriate measurements and settle the issue. Cassini, and France, therefore had the distinction of proving beyond any shadow of doubt that Newton was right.

Meanwhile, in the 1730s Jacques Cassini and his son César François were finally able to go ahead with their national triangulation, financed by the crown. The map they published in 1745 was a triumph, at last accurately fixing the size of France and the relationships between its regions and towns. A thorough topographical survey was still needed, however, if an all-encompassing view of the country was to be obtained. The Cassinis set out to provide this, even finding the means to carry on when royal financing lapsed in 1756, the year of Jacques Cassini's death at the age of seventy-nine. The work also outlasted César François (1714–1784), and the great 182-sheet map was published by the fourth of the Cassinis, Jean Dominique, in 1789.

In 1783 César François also helped to launch the equivalent survey in Britain, persuading the British government to agree to a joint cross-Channel triangulation that would eliminate discrepancies in the values given to the Paris and London observatories. Carried out from 1787 to 1790, the survey was not only valuable in itself, but also paved the way for the creation in the following year of an organization, the Ordnance Survey, that would do for Britain what the Cassinis had done for France.

Below: Louis XV followed the example of previous French monarchs in encouraging the work of mapmakers.

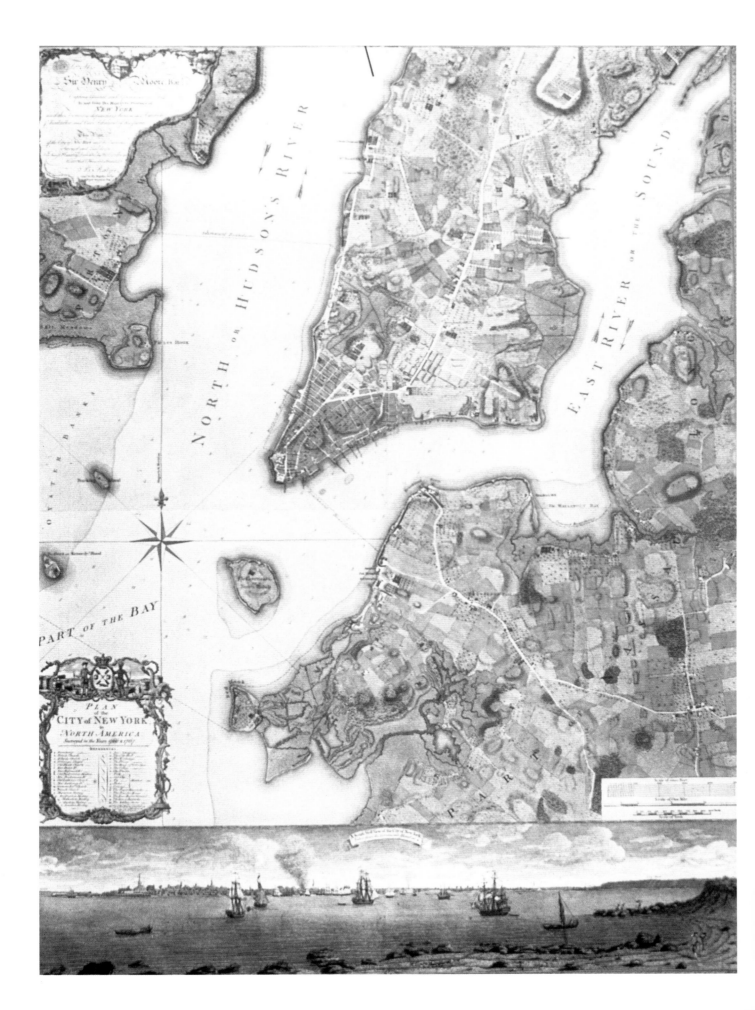

CITY OF NEW YORK
BERNARD RATZER

New York was a small city of some 25,000 inhabitants when a British officer, Lieutenant Bernard Ratzer, surveyed it and drew this highly professional and very decorative map. In the midst of a patchwork of fields and farms, only the built-up area at the tip of Manhattan Island hints at future developments.

Ratzer was a military engineer who was sent to America during the French and Indian War of 1755–1763, which ended in victory for the British and the American colonists. The war prompted thorough surveys of the eastern seaboard, and Ratzer was one of several military men who did the work; among other things, his surveys were responsible for defining the boundaries of New York and New Jersey. The map shows the southern part of Manhattan, with present-day Downtown and the Battery as its core. New Jersey lies to the west, across the Hudson, and Long Island (mostly Brooklyn) lies to the east and southeast, across the East River. Ratzer's map was published in London in two sheets a decade later (1776), by which time a revolt had broken out in the American colonies. Once they had gained their independence, Ratzer's map seems to have been seen as irrelevant to British concerns. At any rate it languished in the Admiralty archives and was rediscovered only recently, its pleasing colors unfaded.

RURAL BROOKLYN
A colorful patchwork of fields occupies the area that is now Brooklyn. Here and elsewhere, the rural nature of most of eighteenth-century New York is obvious.

FAMILIAR GRID PLAN
The severe geometry of these Manhattan streets shows that the island's famous grid plan was already being developed. At the time the map was made, the plan extended as far as what is now 50th Street.

THOROUGH RECORD
A frivolous frame for a sober record of New York's forts, churches, and other important buildings: evidence of the military thoroughness with which Ratzer approached his task.

OLD NEW YORK
The panoramic view of New York is seen from Governor's Island, situated just below Manhattan where the Hudson and East rivers empty into the Upper Bay.

TOWARD INDEPENDENCE

When Bernard Ratzer drew his map of New York City in 1767, the Thirteen Colonies were already riotously discontented with British government. By the time the map was published, only nine years later, the colonies were in a state of rebellion and about to declare for a final separation from the mother country.

After their joint victory over the French, Britain and the Thirteen Colonies soon found themselves at loggerheads. Their divergent interests may well have been irreconcilable in the long run, but British blunders ensured a rapid resort to armed conflict and a permanent separation.

For long periods, Americans had been left to run their own affairs, habitually evading unwelcome regulations. The war was hardly over, however, when a royal proclamation imposed a boundary between colonial and Native American territories that would have prevented any westward expansion across the Appalachians. But it was the British Stamp Act, introduced by Prime Minister George Grenville in 1765, that aroused Americans to riot and boycott British goods. Levied on items such as legal documents and newspapers, the Act was intended to make the colonies contribute to the cost of the war and their continued upkeep. However, the colonists argued that they were not represented in the British parliament and should therefore pay only taxes levied by their own assemblies—the famous cry of "No taxation without representation." In 1766, following Grenville's fall, the Stamp Act was repealed, but a year later new duties, this time on imports, were imposed by the Chancellor of the Exchequer, Charles Townshend. Though less immediately felt, they provoked an equally violent response.

Drift toward rebellion

Matters were not improved by the enforcement of another British measure, the Quartering Act, which required Americans to take a larger share in supporting the 6,000 British troops stationed in the colonies. Once the threat from France had disappeared, the colonists saw no reason to pay, and the troops were popularly viewed as an oppressive standing army, maintained in the interests of tyranny. Friction between colonists and soldiers climaxed in March 1770 in the "Boston Massacre," when rattled Redcoats, surrounded by a hostile crowd, fired and killed five people.

The withdrawal of most of the Townshend Duties had a temporary pacifying effect. Only the imposition on tea was retained, and most colonists simply evaded that by buying smuggled goods. Then controversy arose over trading concessions to the East India Company, and when a consignment of company tea arrived at Boston in December 1773, colonists dressed as Mohawk Native Americans boarded the ships and dumped the tea in the harbor.

This incident, the Boston Tea Party, has become famous, but at the time many colonists were shocked by the wanton destruction of property. The harshness of the British response soon swung opinion around

Above: Enraged by duties levied by a Parliament in which they were not represented, angry colonists throw stamped documents onto a bonfire in Boston to protest the Stamp Act.

Above: Dressed as Native Americans, colonists throw cases of tea into the harbor at the Boston Tea Party, on December 16, 1773. Many colonists were outraged by the demonstration, but were even more alienated by Britain's harsh and clumsy response to it.

again: among other things, Massachusetts' assembly lost most of its powers; a soldier-governor, General Thomas Gage, was installed; and the port of Boston was closed until the destroyed tea was paid for.

Until this time the colonies had never shown much inclination to work together. Now, however, committees of correspondence began to coordinate opposition, and in September 1774 all the colonies except Georgia were represented at the First Continental Congress, where a radical majority condemned "the attempts of a wicked administration to enslave America."

The uncompromising British position reflected a political shift at Westminster. Instead of short-lived administrations, there was now a government with an unchallengeable majority of "King's Friends," led by Lord North and strongly influenced by King George III, whose obstinacy has often been blamed for the loss of the American colonies. British radicals and realists objected unavailingly to the government's policies.

Conflict at Concord

While Massachusetts drifted toward rebellion, the government was slow to reinforce a beleaguered Gage in Boston. Urged to act, in April 1775 Gage sent troops to arrest two radical leaders and destroy an arms dump near the village of Concord. The attempt

to keep the operation a secret failed lamentably, and William Dawes and Paul Revere roused the people of Lexington and Concord. After an encounter with the local militia on Lexington Common, the troops destroyed the dump at Concord, but snipers inflicted heavy losses on them as they retreated to Boston.

The colonies were now unmistakably in revolt. The Second Continental Congress appointed George Washington as commander of the Continental Army, and events began to move rapidly. Rebels led by Ethan Allen and Benedict Arnold captured neglected British forts at Ticonderoga and Crown Point, acquiring valuable artillery. In May 1775 the British won a Pyrrhic victory at Bunker Hill outside Boston, but the city itself was nevertheless besieged and eventually taken by Washington.

Attitudes progressively hardened on both sides. The British declared a naval blockade and hired German mercenaries. In January 1776 the radical thinker Tom Paine published "Common Sense," one of the most influential pamphlets ever written, urging the Americans to make a complete break with Britain.

In June, after a long debate, Congress appointed a committee to draft a declaration of independence, given its final form by Thomas Jefferson and adopted on July 4, 1776.

Rocky Point

Breakers Isle

Gilberts Isles

Parrot I.

Disappointment Cove

Occasional Cove

Pickersgill I.

Goose Cove

Shag River

Cormorant Cove

Passage Point

Petrie Harbour

West Jackel Arm

Pigeon Isles

Sandfell Cove

Duck Cove

Five Fingers Point

Parrot Isles

Five Finger Isles

Bason

Long Sound

Resolution Island

Another Cove

Anchor Island

Seal Rock

Sportsmen Cove

East Point

Cooper's Island

Shag Isles

Facile Harbour

Indian Island

Luncheon Cove

Point

Long Island

Facile Point

Two Front Isles

Cascade Cove

Thrum Cap

Nemesia

Sea Otters

Campbells Harbour

Ship Cove

Goose Cove

The Narrows

A Scale of three leagues

SHIP'S TRACK

Thus appears the North Entrance dist. 2 miles

Thus appears the South Entrance 1¼ mile distant

PICKERSGILLS HARBOUR.

Ship Cove

Watering Place

Astronomers Point

A Scale of 220 fathoms

Cray-fish Island.

The Narrows

A PLAN of

DUSKY BAY

in NEW ZEELAND

PICKERSGILLS HARBOUR

Situated in Lat.ᵈ 45-47-30S

Long.ᵈ 166-08 East

DUSKY BAY
CAPTAIN JAMES COOK

Captain Cook made three great voyages between 1768 and 1779, when he mapped the Pacific and disproved the existence of a southern continent. One of his achievements was a coastal survey of the North and South islands of New Zealand. He described Dusky Bay as "one of the most beautiful which nature unassisted by art could produce."

Cook's rise to eminence was extraordinary in terms of the stratified society of eighteenth-century Britain. The son of a Yorkshire farm laborer, he was apprenticed to the sea at Whitby, where he spent nine years sailing on colliers. In 1756, passing up the chance to become a master, he joined the Royal Navy as an ordinary seaman. Rapidly promoted, he perfected his skills as a navigator and surveyor in the Canadian theater of the Seven Years' War, followed by a five-year stint surveying the coasts of Labrador and Newfoundland. In 1768 the Admiralty gave him the command of the *Endeavour*, a Whitby bark of the type he had often sailed on in his youth, and he set out on his first Pacific voyage. After visiting Tahiti, he sailed to New Zealand, discovered the Cook Strait separating North and South Islands, and surveyed the coasts of both in the course of a 2,400-mile (3,900-km) circumnavigation. In March 1773, on his second voyage, in the *Resolution*, he returned with pleasure to Dusky Bay after 117 days at sea.

ISLANDS GALORE
The details of the chart take on a new meaning for the layperson in a view such as this one, which brings to life the nature of the bay's island-strewn south entrance.

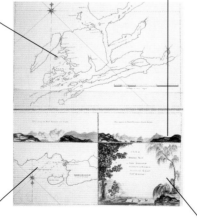

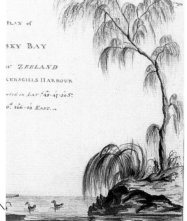

RECENT INVENTIONS
In his meticulously detailed surveying of a difficult, irregular coastline, Cook benefited from the availability of recent inventions and improvements, including the sextant and the station pointer.

HARBOR SOUNDINGS
The chart shows the soundings taken in Pickersgill Harbour—named after one of Cook's colleagues, Richard Pickersgill, said to have been "a good officer and astronomer but liking the grog."

ART
Naturalists and artists accompanied voyages such as Cook's, so expeditions returned with records of great scientific value—and of exotic scenes that fed the eighteenth-century cult of the "noble savage."

COOK AND THE PACIFIC

Captain Cook's meticulous charts reveal the dedication and thoroughness that made him an outstanding, fact-minded explorer. In the course of his three voyages, he disproved centuries-old beliefs and made a host of discoveries, leaving little for his successors to do but tidy up the details.

When Cook set sail in the *Endeavour* from Plymouth, England, in August 1768, there had been only one significant advance in knowledge of the Pacific since Abel Tasman's Australasian voyages in the 1640s. In 1767–1768 a British captain, Samuel Wallis—followed in short order by the French explorer Louis Antoine de Bougainville—rediscovered Tahiti and some other islands that had been found earlier, and then lost again, by Spanish and Dutch mariners.

The transit of Venus

Tahiti was Cook's initial destination. He took with him a group of scientists, an artist to make visual records of everything seen, and the naturalist Joseph Banks,

who, like Cook, would make his reputation as a result of the voyage. Its ostensible (and genuine) object was to observe the transit of Venus across the face of the Sun; Tahiti was one of several locations around the world from which this would be done.

When his public mission was completed, Cook opened his sealed orders, which were to look for Terra Australis, the southern continent, gather further scientific information, and claim any promising territories for the Crown. It was at this point that Cook sailed to New Zealand and spent six months charting its coastline, incidentally establishing that it consisted of two main islands and could not be, as Tasman had believed, part of the southern continent. Then the *Endeavour* sailed up the east coast of Australia (the first ship to do so) and in 1770 anchored in Botany Bay, so called because of its profusion of plant species. Finding the region more congenial than the apparently barren northern and western coasts known to Dutch seafarers, Cook named it New South Wales.

After sailing all the way up the east coast and narrowly escaping disaster on the Great Barrier Reef, the *Endeavour* passed through the Torres Strait (reaffirming Torres' forgotten discovery that New Guinea and Australia were separate) and returned to Britain in July 1771. The wealth of images brought back—of Australian Aborigines, Maoris, Polynesians, kangaroos, and other flora and fauna—revealed to Europe the beauty and strangeness of the Pacific world and made Cook a celebrity.

Cook's second and third voyages

Before long he was on his way again in the *Resolution*, commissioned to find the full truth about the southern continent. He began his second voyage in the summer of 1772 by copying Tasman, approaching the Pacific from the west in order to stay in high latitudes. Over the next three years, the *Resolution* effectively sailed right around the South Polar region, farther south than any ship under canvas had ever done. It crossed the Antarctic Circle three times and penetrated as far south as latitude 71 degrees.

Left: James Cook rose from humble origins to command, explore, and set new standards in navigation and surveying.

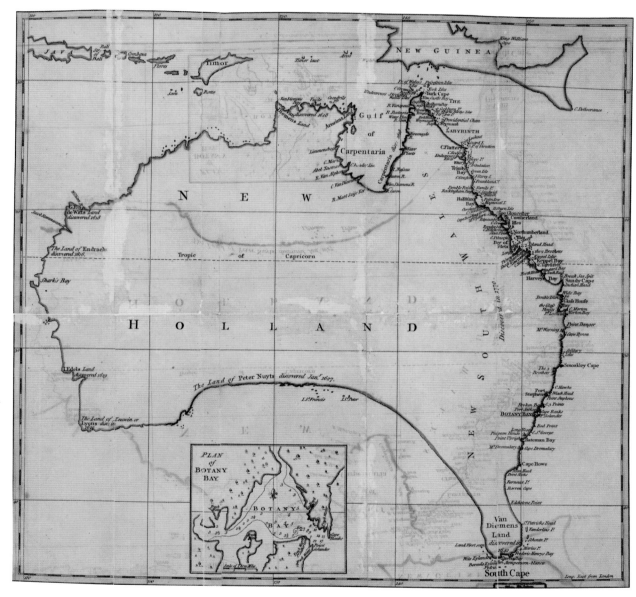

Above: A map of New Holland from the 1770s shows New South Wales and an inset of Botany Bay, named by Cook for its wildlife. Original plans for a convict settlement at the site were abandoned in favor of nearby Port Jackson, later the site of Sydney.

The vast and wealthy continent of Terra Australis was shown to be a fable; if any land at all existed in the far south, it was buried beneath impassable ice. In the course of his voyages, Cook visited Tierra del Fuego, South Georgia, Tonga, New Caledonia, the Marquesas, and Easter Island. Now equipped with John Harrison's chronometer, which made it possible to establish longitude at sea, Cook could record islands' locations accurately for the first time.

In 1776 Cook set out again to test another long-held belief, in the existence of a northwest passage from the Atlantic to the Pacific. Two centuries earlier British seamen had sought to find it via Hudson Bay. Cook tried the other, Pacific side, investigating every inlet on the Pacific coast of Canada and Alaska, passing through the Bering Strait and across the Arctic Circle. Once more he persevered until driven back by the ice. During his explorations he had discovered Christmas Island and the Hawaiian group. Returning to Kealakekua Bay on Hawaii, Cook became embroiled in a dispute with natives over a stolen boat, and on February 14, 1779, he was stabbed to death in a scuffle on the shore. However, his work—the mapping of the Pacific—was largely done.

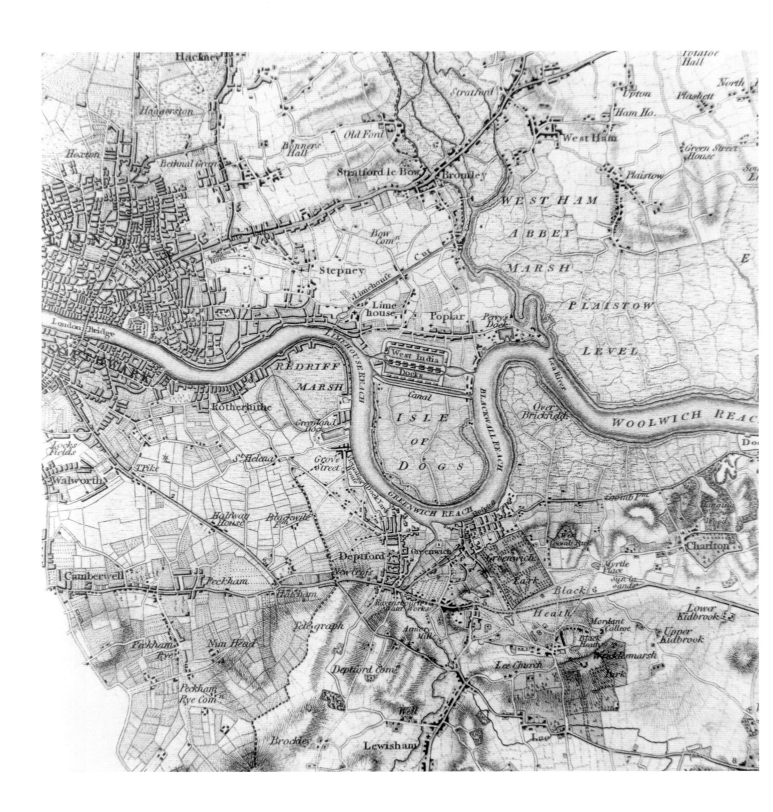

THE THAMES RIVER
ORDNANCE SURVEY

The first sustained effort to map Great Britain in its entirety began in 1791. The results were published from 1801 by an organization that became known as the Ordnance Survey. Like the similar effort by the Cassinis in France, it marked the beginning of mapping as an ongoing national enterprise.

The first Ordnance Survey production was a four-sheet map of Kent, published in 1801. Though the Ordnance Survey was to become a major publisher in its own right, it began by using the services of an established commercial supplier of maps to the Crown, William Fadan. The map of Kent was extended to take in southern Essex, just along the Thames, so that the width of the river could be included on the map. The area shown in this detail had not yet been swallowed up by the eastward spread of London. Small settlements and open fields are much in evidence, but the land south of the river was marshy and insalubrious, with run-down wharves and steps inhabited by the disreputable poor until a more spacious development was reached at Greenwich. When the map was published, the West India Docks were newly constructed. They proved to be the first of a series of vast modern docks that appeared within the next few years to accommodate the shipping and commerce of the booming Port of London.

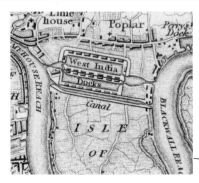

AWE-INSPIRING DOCKS
West India Docks, opened in 1802, were described by a German visitor as "an immeasurable work, at seeing which the most cold-blooded spectator must feel astonishment, and…awe at the might of England."

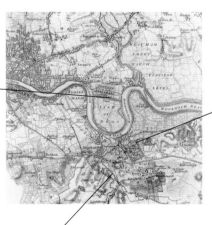

ROYAL PARK
Greenwich Park was laid out for King Charles II and was the site of the famous Royal Observatory, established in 1675. The zero meridian of longitude passes through the former observatory (now Flamsteed House).

SHADING FOR HEIGHT
Hachures have replaced profile drawings as indicators of high ground. Contour lines would come into common use only later in the nineteenth century.

A NATIONAL INSTITUTION

The enthusiasm of Major General William Roy, combined with a sense that Britain was falling behind its continental neighbors, led to the founding of an organization that would map the British Isles. Over the two centuries of its existence, the Ordnance Survey became a national institution.

Above: Charles Lennox, duke of Richmond, who, as master of ordnance, founded the Trigonometric Survey.

The need for reliable, state-sponsored surveys and maps first became apparent in Britain during the Jacobite Rebellion of 1745–1746. Though successful in defeating Bonnie Prince Charlie's Highlanders, the Duke of Cumberland's army was hampered by misinformation about the nature of the terrain, and Cumberland himself complained to his father, King George III. As a result, in 1747 a survey of the Scottish Highlands was undertaken, which was later extended to the whole of Scotland. It was more or less completed, although the imminence of war with France brought operations to an abrupt end in 1755. Much of the work was carried out by Scottish enthusiast William Roy, and the Military Survey of Scotland is often informally referred to as "Roy's Map."

Roy was already convinced that a national British survey was needed, but the proposals he put forward over the years were rejected as being far too costly. The project was revived only in 1783, when the French geographer César François Cassini proposed a linked Anglo-French survey that would determine the relative positions of the Paris and London (Greenwich) observatories. The British government agreed, and Roy took charge of the triangulation of southeast Britain. Public interest was aroused by the sight of soldiers on Hounslow Heath using specially manufactured glass tubes to measure the baseline, and both the French techniques and British technology—the glass tubes and the advanced theodolite constructed by Jesse Ramsden—impressed observers.

The success of the survey strengthened Roy's case, although he died in 1790, just before his project was at last taken up. In June 1791 the Duke of Richmond,

Master of Ordnance, paid Jesse Ramsden £373.14s. for a 3-ft. (90 cm) theodolite, a transaction generally regarded as the first piece of business done by the Trigonometric Survey of the Board of Ordnance, later known simply as the Ordnance Survey. Thus the national survey began under the auspices of a military organization, and the tradition of military control remained strong until the first civilian director was appointed in 1977.

The first Ordnance Survey map
The board's first objective was to produce a map of the entire country on a scale of 1 in. (2.5 cm) to 1 mi. (1.6 km). This was a familiar scale, which had

already been used on a number of eighteenth-century county maps. (In 1858 a Royal Commission would confirm the 1-in. scale as a standard for British mapmaking, along with 6 in. and 25 in.) In 1795 a 1-in. (2.5 cm) map of Sussex was published. It was the work of a surveyor and an engraver who were funded by the duke and who had been taken on to the Ordnance Survey staff. Officially, however, the first Ordnance Survey production was the four-sheet map of Kent published by William Fadan in 1801. It was followed in 1805 by a map of neighboring Essex, which effectively marked the emergence of the Ordnance Survey as a self-sufficient organization— engraving and publishing, as well as surveying and drawing, its own maps. The title "Ordnance Survey" was used for the first time in 1810 on a map of the Isle of Wight and part of Hampshire.

Clear and simple maps

These early Ordnance Survey maps were models of clarity, but they were black and white, contained no standard symbols, and employed only hachures, or shadings, to indicate high ground—contour lines only replaced hachures from 1839–1840. The first maps in full color were not published until 1897. The national survey was not completed until 1870, thanks to a range of distractions including wars; reductions in funding; separate surveys of Ireland— on the 6-in. scale—and Jerusalem; and protracted disputes as to which scales were most suitable for rural, urban, and other maps. Despite the delays, the Ordnance Survey established its superiority over its private competitors, beginning a long and distinguished career that has extended into the age of computers and digital mapping.

Above: Charles Stewart, "Bonnie Prince Charlie," conquered Scotland and defeated government troops at the Battle of Prestonpans in 1745. Later, the difficulty of tracking down Charles's Jacobites helped to bring about the founding of the Ordnance Survey.

A
Map of
LEWIS AND CLARK'S TRACK,
Across the Western Portion of
North America
from the
MISSISSIPPI TO THE PACIFIC OCEAN:
By Order of the Executive
of the
UNITED STATES,
in 1804, 5 & 6.

NORTH AMERICA
LEWIS AND CLARK'S EXPEDITION

The American West was largely unexplored and unmapped until 1804–1806, when Lewis and Clark made their epic journey from St. Louis to the Pacific. Where they led, many thousands were to follow, along the Oregon Trail, to settle the land.

By the end of the eighteenth century Americans had realized that vast areas existed to the west of the United States, with the Rocky Mountains forming a barrier stretching down the continent. Thomas Jefferson, who became president in 1801, was anxious to explore, and claim, the West, and also to find a swift river passage to the Pacific. He chose his secretary, Meriwether Lewis, to lead an expedition, and Lewis coopted a fellow army officer, William Clark, as his partner. By the time they set off, the motive for exploration had become even stronger, since Jefferson had concluded the Louisiana Purchase, acquiring from France the undisputed ownership of the territories between the Mississippi River and the Rockies. Expansion beyond the mountains, and the creation of a transcontinental nation, beckoned. Lewis and Clark's expedition was the first to trace the Missouri River to its source, after which it found a way through the Rockies and reached the Pacific. The swift river route across the continent proved to be a chimera, but Clark's maps showed the way for the trappers, adventurers, and settlers who followed.

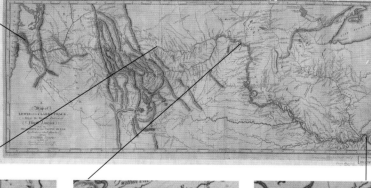

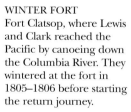

WINTER FORT
Fort Clatsop, where Lewis and Clark reached the Pacific by canoeing down the Columbia River. They wintered at the fort in 1805–1806 before starting the return journey.

MIGHTY FALLS
The Great Falls of Missouri, 100 ft. (30 m) high, forced the expedition to make carts and drag their boats and equipment round it, a portage that took more than three weeks.

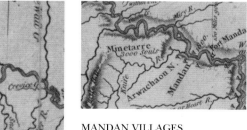

MANDAN VILLAGES
The Mandan Native American villages where Lewis and Clark wintered in 1804–1805 before venturing into entirely unknown territory. Here the Shoshoni Sacajawea joined the expedition, contributing greatly to its success.

STARTING POINT
St. Louis, at the junction of the Mississippi and Missouri rivers. Lewis and Clark left on the Wood River, close to the city, and began their long upriver journey on May 14, 1804.

CROSSING THE CONTINENT

In two and a half years, Lewis and Clark's party traveled more than 7,400 miles (12,000 km) through Native American territory, often stretched to the limit by the climate or terrain. Well prepared and well led, the expedition lost only one man and achieved all of its objectives.

In the winter of 1803–1804 Lewis and Clark assembled and trained a "Corps of Discovery" at St. Louis, consisting of about forty soldiers and frontiersmen. The two leaders were themselves formidable men in their thirties with experience of the frontier and Native American fighting. Neither was formally well educated, but Lewis was a careful observer of flora and fauna, and Clark proved to be a self-taught mapmaker of impressive skill. Their journals, though filled with comic misspellings, are deeply absorbing narratives. Lewis and Clark got on well together and maintained their authority over their men, on occasion by harsh disciplinary measures. They also had the advantage of being sponsored by

the U.S. government, using funds from Congress to equip themselves and acquire quantities of gifts for the Native American peoples they would encounter. In the event, relations with the Native Americans were remarkably warm, and on occasion the party would probably not have survived without their assistance.

Setting off from St. Louis

On May 14, 1804, Lewis and Clark and their men left St. Louis in a barge and two small sailing boats (pirogues). For five months they followed the Missouri upriver, passing peacefully through Sioux country but enduring fierce attacks from mosquitoes. They wintered at the Mandan villages, where they acquired a French–Canadian guide named Toussaint Charbonneau and—as it turned out, even more importantly—his pregnant Shoshoni wife, Sacajawea. She became guide, interpreter, and intermediary when the party encountered her own people.

The expedition had so far traversed country of which they had some knowledge. In April 1805 they set off in smaller boats into completely unknown

Above: In this illustration from an account of the expedition by Peter Gass, Lewis and Clark hold a council with a group of Native Americans. Although relations with the native inhabitants were generally good, the party did sometimes find itself under attack.

Above: Sunlight through a hole cut in the roof illuminates the dwelling of a Mandan chief, illustrated in the mid-nineteenth century. The Lewis and Clark expedition wintered in the Mandan villages, where it acquired an essential guide, Sacajawea.

territory, traveling steadily westward along the Missouri. When they reached the Great Falls of the Missouri, they were forced to make a portage, constructing carts to haul their boats overland until they could safely take to the river again. It led them to Three Forks, where three streams met to form the Missouri. They followed the westernmost (the Jefferson River) until they located the Lemki Pass, south of the Bitteroot Mountains.

Crossing the Rockies

They were now in the heart of the Rocky Mountains. The boats were abandoned and the party marched north until they made contact with the Shoshoni. Thanks to Sacajawea, the Shoshoni supplied them with food, horses, and information. After a few weeks the expedition marched on through the Lolo Pass, left the Rockies behind, and reached the Clearwater River. Having constructed dugout canoes, they made the long and often dangerous journey along the Clearway and Snake rivers, and into the great Columbia River, which carried them all the way down to the Pacific.

On November 15 they arrived at the mouth of the river, where they built Fort Clatsop and spent the winter. Lewis and Clark expected that the entire party would be taken off by a passing ship, but when none appeared by March, they decided to return the way they came. They made a few exploratory diversions,

notably when Clark left the main party for a while to investigate the Yellowstone River region. Peaceful relations with the Native Americans proved harder to maintain, and two Blackfoot were killed in a skirmish that turned the tribe into enemies of the whites.

Finally, after a six-month trek, the expedition reached St. Louis on September 23, 1806, having recorded, mapped, and opened up vast areas of the North American continent.

Above: Rainbow Falls, Montana, part of the Great Falls of the Missouri that forced Lewis and Clark and their party to make a diversion involving an arduous three-week portage.

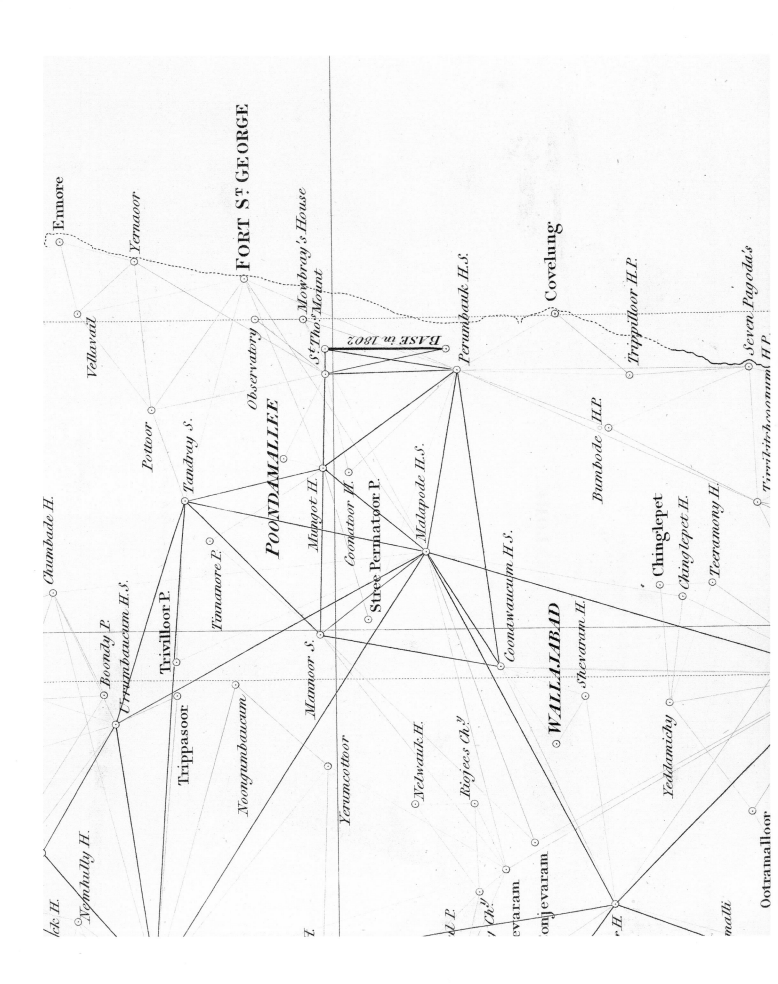

GREAT SURVEY OF INDIA
WILLIAM LAMBTON

The British East India Company began trading on the west coast of India in the early seventeenth century, and evolved into a military and political power whose influence was felt throughout the subcontinent. The company's commitments made reliable maps a necessity, and doughty British and Indian surveyors were at work from the late eighteenth century.

The most all-embracing survey was proposed by Captain William Lambton, an army officer who convinced the company's directors to undertake a complete and exact survey of all parts of India controlled or influenced by the company. The method was the well-tried one of triangulation. A measured base line served as one side of an imaginary triangle, a visible distant point was selected, and a theodolite was used to measure the angle between each end of the base line and the sightline linking it to the distant point. The length of the two lines was then calculated by trigonometry, and each could be made the base for a new triangle, and so on. A straightforward process in principle, triangulation proved laborious and life-threatening when applied on a continental scale, over terrain ranging from desert to forest and mountains and impeded by mists, fevers, and local sensitivities. The work, carried on by Lambton, George Everest, and their successors, was a heroic affair, ultimately known as the Great Indian Trigonometrical Survey.

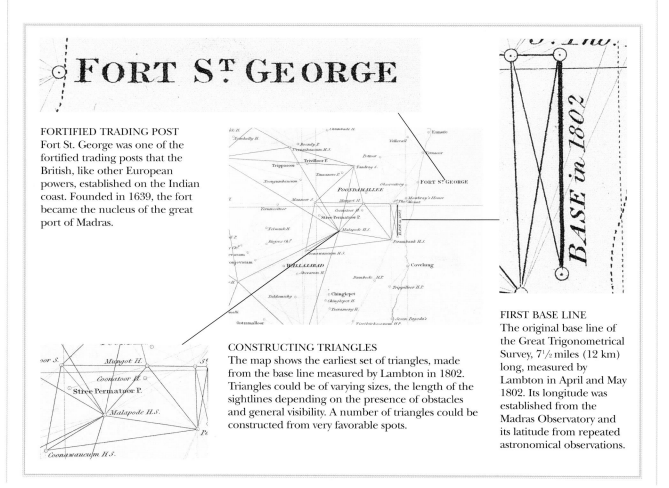

FORTIFIED TRADING POST
Fort St. George was one of the fortified trading posts that the British, like other European powers, established on the Indian coast. Founded in 1639, the fort became the nucleus of the great port of Madras.

CONSTRUCTING TRIANGLES
The map shows the earliest set of triangles, made from the base line measured by Lambton in 1802. Triangles could be of varying sizes, the length of the sightlines depending on the presence of obstacles and general visibility. A number of triangles could be constructed from very favorable spots.

FIRST BASE LINE
The original base line of the Great Trigonometrical Survey, 7½ miles (12 km) long, measured by Lambton in April and May 1802. Its longitude was established from the Madras Observatory and its latitude from repeated astronomical observations.

TRIANGULATION OF THE GREAT ARC

The triangulation of India was an epic achievement, costly in lives as well as money. Its central spine was the Great Arc of the Meridian, a series of interlocked triangles stretching the entire 1,600-mile (2,570-km) length of the Indian subcontinent.

Lieutenant William Lambton reached India in 1796 after years of routine service in New Brunswick. During that time he had made himself an expert in surveying and geodesy. He took part in the Fourth Mysore War, which extended the East India Company's territory into southern India—a propitious moment at which to propose an ambitious survey that would map British India and also represent a large step toward solving the question of the exact curvature of the Earth.

Lambton starts the survey
Though an awkward and self-effacing personality, Lambton had impressed his superiors and in 1800 was commissioned to carry out the survey. Lack of adequate instruments delayed the project, and although Lambton managed to acquire a reliable steel chain with which to measure out the base line, the half-ton Great Theodolite he had ordered was unaccountably delayed. It transpired that the ship carrying it had been captured by the French, who eventually courteously forwarded the theodolite from Mauritius. Meanwhile, in April 1802 Lambton spent fifty-seven days laying out the base line, a process that involved making 400 measurements. Such pains were typical of his

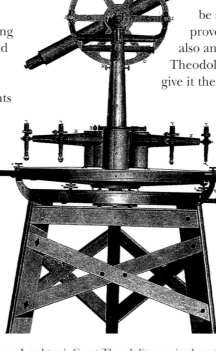

Above: Lambton's Great Theodolite survived capture by the French and the necessity of impromptu repairs to give years of field service.

approach: He insisted on measuring all three angles of his triangles, and many observations and calculations were repeated and revised again and again, ensuring an extraordinary level of accuracy.

Lambton's first major effort was not the Great Arc but a coast-to-coast triangulation of the peninsula (Madras to Mangalore) in 1803–1805 that served to advertise the usefulness of his work: The distance was found to be 360 miles (580 km) instead of the then-accepted figure of 400 miles (645 km). After this the triangulation of the Great Arc went forward, at first mainly south until it reached the tip of the peninsula, and then steadily north.

The fieldwork was undertaken during and just after the monsoon season, which at intervals offered better visibility (dust- and haze-free) than at other times, along with frequently atrocious, fever-breeding conditions. In addition Lambton had to cope with local suspicions, especially strong on the part of dignitaries who believed that Englishmen looking through lenses from high places must be spying on their womenfolk. Lambton proved to be an excellent diplomat and also an indomitable chief: When the Great Theodolite, hoisted on to a temple gateway to give it the necessary height for readings, crashed to the earth, he spent six solid weeks in a tent, pulling and pegging and hammering until the buckled dial had been mended.

By 1817 the scale of Lambton's achievement was acknowledged, and his project became officially known as the Great Trigonometrical Survey of India. Bombay and Calcutta were to be locked into the arc by branching chains of triangles, and the arc itself was to extend to Agra. In 1823, however, still pressing ahead, Lambton suddenly died.

Lambton's successor was thirty-two-year-old George Everest, who had joined the

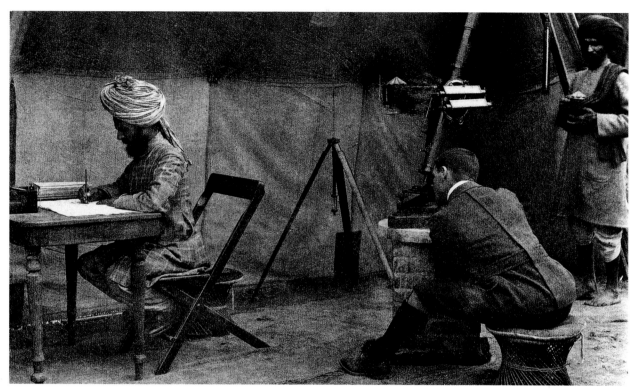

Above: George Everest, with his back to the camera, reads out measurements to a Sikh assistant during the Great Survey.

survey in 1818. His first season had ended in disaster, when his entire party of 150 were laid low with fever and fifteen had died. From that time Everest suffered from appalling health, and his obsessive drive to complete the arc often seemed like a race against death.

Very different from Lambton, Everest resembled the stereotypical Indian Army officer, choleric and quarrelsome. He was responsible for important innovations, though, notably working at night with lamps, and later flares, which proved to be visible even through mists. This meant that the work could be undertaken during the healthier dry seasons instead of during the monsoon.

The arc completed

Everest drove the arc through central India and then, sick again, left for England while his assistants extended the triangles from the arc to Calcutta in the east. After an absence of five years, he completed the link with Calcutta, where the flat country and dense

vegetation meant that there was a lack of vantage points. Some telegraph towers offered the required elevation, but Everest also built towers to fill the gaps.

This successful, if expensive, experiment became the basis for Everest's plan to push on beyond Agra: Building towers would make it possible to survey the great plain to the north as far as the approaches to the Himalayas and the limits of British India. By this means, despite immense difficulties, the Great Arc was finally completed in 1837.

Everest, once more desperately sick, retired in 1843. On his return to England, his health improved dramatically and he married, fathered six children, was knighted, and died in 1866 at the age of seventy-six.

Meanwhile, further surveys were undertaken, including one along the edge of the Himalayas. In 1856, when "Peak XV" seemed clearly established as the highest in the world, Everest's successor, A. S. Waugh, proposed that it should be called Mount Everest—a name that, despite objections, has stuck.

GEOLOGICAL MAP OF THE WORLD.

ICE BARRIERS OF THE POLAR REGIONS

The north Polar Regions consist chiefly of primitive and transition rocks, with few secondary and alluvial and slight tertiary strata. Coal of the oldest formation was found at Melville Island, and the plants of the coal formations of Baffins Bay are similar to those which now flourish between the tropics.

Coral reefs are the work of organic beings which exist in inappreciable numbers. They consist of agglutinated skeletons of departed races of polyp, composed of carbonate of lime, cemented into hard calcareous rock.

CORAL REEFS

Spitzbergen

Iceland

ATLANTIC

Azores

Canary I.s

Cape Verde I.s

Sierra Leone

Timbuctoo

Lisbon

Elsinburgh

Berlin

Petersburg

Moscow

Bombay

Constantinople

Cairo

AFRICA

St Helena

Cape Town

Madagascar

INDIAN

OCEAN

CORAL

Is of
Japan

PACIFIC

Philippine

REEFS

OCEAN

AUSTRALIA

Portland

Sidney

New Zealand

REGION

PACIFIC

OCEAN

SOUTH AMERICA

Buenos Ayres

Rio Janeiro

Lima

West India I.s

Bermudas

New York

New Orleans

San Francisco

Nova Scotia

NORTH AMERICA

Drawn & Engraved

by John Emslie.

Barren Island, Bay of Bengal.
One of the most remarkable volcanic islands now in action. The cone emits vast volumes of smoke and red hot stones: some, of which weigh three and four tons.

REFERENCE

4	ALLUVIUM	Sand, Gravel.
3	TERTIARY	Drift, Cray, Clay.
2	SECONDARY	UPPER: Chalk, Oolite, Red Sandstone
		LOWER: Coal, Limestone, Devonian.
1	PRIMARY	Mica, Gneiss, Quartz, Granite
	VOLCANIC ROCKS	Trap, Greenstone, Porphyry.

(coral Reefs) *This mark shews the localities where Coal has been found.

Published by James Reynolds, 174 Strand London.

Island of Cyclops Mediterranean.
A volcanic formation of tuff clay, and associated lava.

Cape Pillar, Van Dieman's Land.
A remarkable basaltic formation.

GEOLOGICAL WORLD MAP
JAMES REYNOLDS

In the nineteenth century geology began to be an exact science based on chemical analysis of rocks and empirical investigation undertaken in the field. Geological maps became common after 1815, but James Reynolds' publication, though decorative, could claim to be no more than a provisional account of world geology.

Though geology was a young science, by the time James Reynolds' map appeared in 1849, many parts of the Western world had been surveyed and recorded with scientific rigor. Reynolds used these findings but was evidently unwilling to admit that anything remained to be found out. His map was presented as though it was comprehensive, even though much of the interior of Africa and Asia had not yet been explored, let alone subjected to geological analysis. In fact, the kind of map that

Reynolds had endeavored to produce became possible only around 1900, after a half-century of large-scale surveys and some heroic feats of exploration. However, his Geological Map of the World is interesting precisely because it records the current state of knowledge, including its deficiencies and delusions, while also providing the reader with an enjoyable visual experience. It is particularly noteworthy as a late example of a genre that was dying by the mid-nineteenth century—the illustrated map.

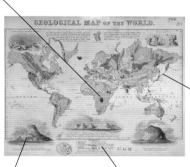

SIMPLIFIED AFRICA
The geology of southern Africa is shown in very simplified form. In addition to the mainly "Primary" and lower "Secondary" levels represented here, the region has a much more complex configuration of later deposits.

CHAIN
A chain of volcanoes stretches around the Pacific Rim, one section passing through the Philippines and Japan and across the north Pacific. This "ring of fire" is caused by the collision of two of the gigantic plates that make up the Earth's crust.

CYCLOPS ISLAND
The volcanic island of Vulcano is one of the Lipari Islands, off the north coast of Sicily. In classical mythology one-eyed giants, the Cyclopes, worked in a subterranean forge on Vulcano, making thunderbolts for the god Jupiter.

KEY
Reynolds' key is arranged as a fourfold geological sequence. His "Primary" and "Secondary" are now known as the Paleozoic and Mesozoic. "Alluvium" refers to the most recent geological phase, the Quaternary, identified in 1829.

EARTHLY DISPUTES

During the late eighteenth and nineteenth centuries, geology developed into a scientific discipline. During its early days, however, there were a number of fierce controversies over fundamentals: above all, over how rocks were formed and how old the Earth could possibly be.

These were sensitive subjects because they impinged on accepted religious views concerning the Creation and early history. In the late eighteenth century, a lecturer at the Freiburg Mining Academy, Abraham Gottlob Werner, became the principal champion of the neptunist theory, so called from Neptune, the Roman god of the oceans. Werner held that the Earth had been entirely covered by primeval waters and that, as they subsided, layers of rocks were deposited by precipitation, starting with igneous materials such as basalt and granite.

One of the advantages of the neptunist theory was that the primeval waters could be equated with the biblical Flood. The rival plutonist view was associated with the Scottish geologist James Hutton, often regarded as the founder of geology as a science. Hutton emphasized the role of the heat and pressures deep within the Earth (Pluto was the god of the underworld) in creating igneous rocks. He was able to provide examples of granite forming fingers and veins within supposedly more recent rocks, or even surrounding them completely—a situation incompatible with the supposed origin of all granite as a chemical precipitate forming the oldest rocks.

Hutton's ideas were rather obscurely expressed, and they became well known only after his death, when his friend John Playfair published *Illustrations of the Huttonian Theory of the Earth* (1802). Evidence against the neptunist theory accumulated, and by the time of Werner's death in 1817, it had effectively been discredited.

Above: The French natural scientist Georges Cuvier, seen here lecturing on paleontology at a Paris museum. He was the leader of the "catastrophist" school, proposing what now seems a ludicrously brief age for the Earth of 75,000 years.

Above: A volcano erupts in Hawaii. The uniformitarians believed that such geological processes had shaped the Earth over a period of millions of years, in which case the planet was far older than catastrophists believed.

A closely related controversy pitched the catastrophists against the uniformitarians. The catastrophists interpreted phenomena such as folded and tilted strata as evidence of sudden upheavals—"catastrophic" events that had altered the Earth's history, for example, by causing mass extinctions of species. One corollary of catastrophism was that known developments could have occurred within a very short time, and the chief of the school, the great French natural scientist Georges Cuvier, put the age of the Earth at a mere 75,000 years.

The ascendency of uniformitarianism

The opposing doctrine was implicit in James Hutton's writings. These proposed an immeasurably long Earth history in which gradual changes over the ages produced tremendous effects—geological cycles of erosion, volcanic activity, deposition, subsidence, and uplift. All of these were processes that were still taking place, so that it was possible for the geologist to study the past through the present. Uniformitarianism was embraced by another Scottish geologist, Charles Lyell, whose *Principals of Geology* (1830–1833) became a classic authority. Though uniformitarianism became the received doctrine, some of its more rigid attitudes have since been abandoned (for example, refusal to acknowledge fluctuations in volcanic activity), and the impact of some catastrophic elements (such as meteorite impacts) has been acknowledged.

As a result of Hutton and Lyell's work, it became accepted that the Earth was much older than had previously been believed. But the first scientific efforts to date its age evoked surprise and disagreements. In 1860 John Phillips made a calculation that was based on the rate at which sedimentation was believed to occur. He concluded that it must have taken more than ninety-five million years to create existing strata. Two years later his estimate appeared to have been confirmed when one of the nineteenth century's greatest scientists, the physicist William Thomson (Lord Kelvin), made a separate calculation, based on the cooling of the Earth, and arrived at a figure of a hundred million years.

Some geologists became uneasy about the smallness of these numbers. The biologist Charles Darwin was even more concerned, since evolution as he had described it could not have taken place in such a short time. Then, in the early years of the twentieth century, Ernest Rutherford demonstrated that the operation of a newly discovered heat source—radioactivity—vastly extended the age of the Earth. Thomson's estimate was discredited—rightly, although it later transpired that the mode of cooling (convection not, as Thomson supposed, conduction) was more significant than radioactivity. Subsequent refinements of method and understanding have led to the current figure for the age of the Earth: about 4,550 million years.

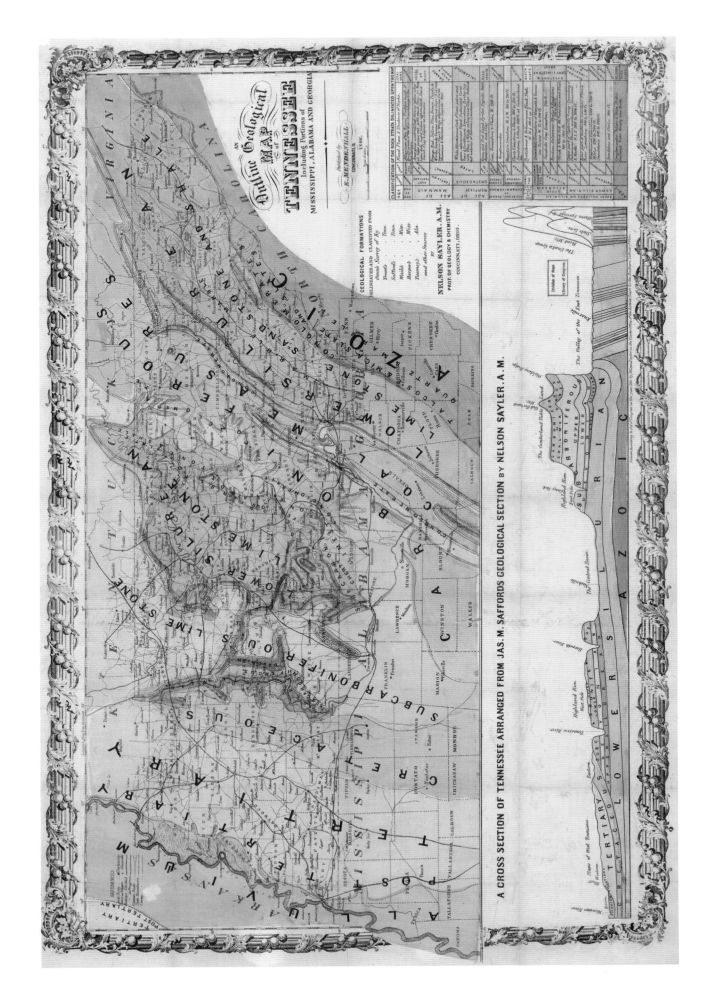

TENNESSEE
NELSON SAYLER

With a vast new country opening up, Americans were eager to record its topography and natural resources, including those hidden beneath the surface of the Earth. The American Geological Society was founded in 1819, and within a few years the first surveys were mapping the rock strata of the individual states.

Sayler's map of Tennessee was published in 1866, but it derived from the findings of surveys earlier in the century. The first American survey, of South Carolina, dated back to 1824, and activity intensified in the 1830s as states began to set up their own geological departments. Their motives were practical, as well as scientific, since the identification of rock types constituted a first step toward locating valuable resources such as coal and metals. Federal-sponsored activity came to a climax with the "Great Surveys" conducted between 1867 and 1879—scientific expeditions ranging over a huge area of the West, headed by the geologists Ferdinand V. Hayden, Clarence King, and John Wesley Powell. Among their feats were Powell's journey down the Colorado River through the Grand Canyon and Hayden's exploration of the upper Yellowstone region, leading to the establishment of the Yellowstone National Park. A single body, the U.S. Geological Survey, was founded in 1879 and has remained impressively active ever since.

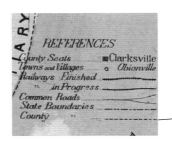

SUBSTRATUM
Underlying the color-coding and boldly printed geological labels is a more familiar map of Tennessee and parts of neighboring states, with boundaries, cities, railroads, and other features indicated by conventional symbols.

TABLE
The rock strata are very comprehensively and neatly classified and color-coded. Columns are devoted to their age, period, epoch, and common name, along with a description of their type, place, and thickness.

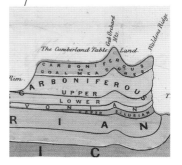

COAL
The link between geology and the quest for mineral resources is highlighted here by the switch from technical language (carboniferous) to the ordinary language identifying this as the region of coal measures.

CROSS-SECTION
The Cumberland Plateau. Safford's cross-section illustrates the geology underlying the topography of Tennessee, from the Unaka Mountains in the east to the Mississippi alluvial plain on the western edge of the state.

RECORDING THE ROCKS

Maps showing the distribution of rocks on and beneath the surface of the Earth are now familiar objects, but the evidence on which they are based was gathered laboriously over the generations. In Europe, America, and the colonial empires, government-sponsored surveys played a large part.

Priorities are often difficult to establish in science, where, for example, a conception may be voiced at a very early date, acquiring substance over a long period and by stages—any one of which could be described as its "real" scientific beginning. Hints of geological insight are found in antiquity, when some thinkers argued that the presence of fossilized shells on land must indicate violent past upheavals within the Earth. Mineralogical lore accumulated from the fifteenth century as mining became an important economic activity. Leonardo da Vinci made remarkable drawings of rock strata, and in 1669 the Danish physician Nicolaus Steno proposed that such strata were laid down in a chronological sequence, with the most recent at the top. Shortly afterward a seventeenth-century writer, Martin Lister,

proposed the making of colored or shaded maps of soils. And in 1746 a Frenchman, Philippe Buache, made what are claimed to be the first actual maps of Western European rock formations and minerals, complete with shading and specific symbols.

However, when all necessary acknowledgments have been made, William Smith remains the key figure in both geological mapping and the wider aspects of the subject. He has rightly been hailed as "the father of English geology"—all the more aptly in that most of the other British pioneers were Scots.

Smith discovers a system of rock chronology

Smith was a civil engineer who worked in various parts of the country as a surveyor for canal-building projects. Investigating the rock outcrops he encountered, he noticed that certain animal and vegetable fossils were always found in the same stratum of rock, and that this remained the case at widely different locations. His observations also confirmed the belief that strata were always deposited with the oldest at the bottom and the youngest on top, the fossil deposits evolving in parallel from relatively primitive to more advanced forms. This meant that a chronological sequence of strata could be constructed, and that strata could be identified on the basis of the fossils they contained.

Left: With the understanding of stratification developed by William Smith and others, it became possible to attempt ever more ambitious engineering projects, such as the building of the Panama Canal, pictured here in 1898.

Above: A trilobite fossil dating from the Cambrian period, 570 million years ago. Smith realized that strata contained similar fossils, suggesting that they were contemporaneous.

Smith published some of his findings in *The Order of the Strata and Their Embedded Organic Remains*, which made a notable contribution to geology and the related disciplines of stratigraphy and paleontology. Meanwhile he prepared his twenty-four-color *Geological Map of England and Wales, with Part of Scotland*, finally published in 1815.

All of Smith's achievements were accomplished in the face of severe personal and financial difficulties that kept him working as a surveyor while struggling to complete his scientific projects. Before the publication of his map, he was forced to sell his collection of geological specimens; after its publication, he was imprisoned for debt. Somehow, in 1822, he managed to publish a set of more detailed county and sectional maps, his *Geological Atlas of England and Wales*. His work was not ignored: It was widely employed to confirm stratigraphic sequences elsewhere that became the basis of new classifications, and on occasion his findings were simply plagiarized. Personal recognition was long delayed, and Smith's great contributions to science were recognized only in the decade before his death in 1839.

The establishment of national surveys

Though Smith's work was remarkably accurate, new studies were carried out from 1835, when Britain became the first country to establish an official national survey. Meanwhile, the geology of France was mapped with exceptional thoroughness by Jean-Baptiste de Beaumont, although a national survey was not established until 1867.

Other European countries followed suit, while also exploring and mapping their colonies. W. E. Logan's Canadian survey, carried out in 1865–1869, was one of many British imperial operations. In the United States decades of strenuous surveying—by expeditions and surveys for the railroads—culminated in the setting up of a national geological survey in 1879. By the time the 1900 International Exhibition opened, it was possible to display convincingly detailed geological maps covering the entire world. Nevertheless, much remained to be done in the following century, including the mapping of the world's ocean floors.

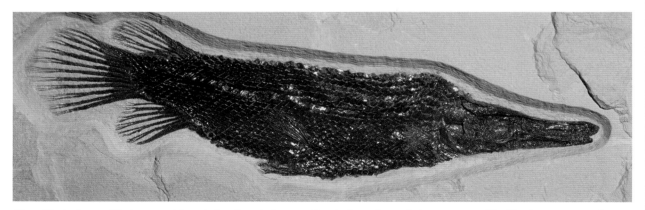

Above: Geologists rapidly became able to judge the relative age of rocks from the evolutionary stage of fossilized creatures, such as this fish, found within them. It only became possible to date rocks absolutely in the twentieth century.

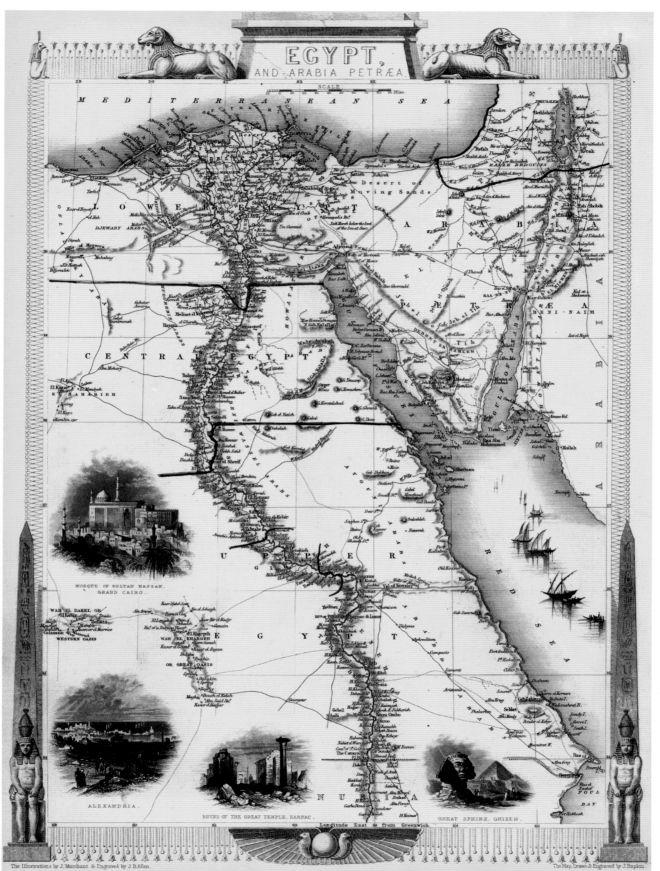

EGYPT,
AND ARABIA PETRÆA.

SCALE

MEDITERRANEAN SEA

LOWER EGYPT

DJEWABY ARABS

CAIRO

CENTRAL EGYPT

ET BAHARIEH

Desert of Moving Sands

ARABIA

HAZER BEDOUINS

BENI-SAIM

RAA-ES

UPPER

MOSQUE OF SULTAN HASSAN.
GRAND CAIRO.

WAH EL DAKEL OR
WESTERN OASIS

WAH EL KHARGEH
OR GREAT OASIS

EGYPT

RED SEA

GULF OR ARABIA

ALEXANDRIA.

RUINS OF THE GREAT TEMPLE, KARNAC.

NUBIA

ARABIA

POUL
BAY

GREAT SPHINX, GHIZEH.

Longitude East 33 from Greenwich.

The Illustrations by J. Marchant. & Engraved by J.B.Allen.

The Map, Drawn & Engraved by J.Rapkin.

626888

J. & F. TALLIS, LONDON, EDINBURGH & DUBLIN.

EGYPT
JOHN TALLIS

During the nineteenth century British maps were researched and drawn with increasing scientific rigor and engraved with astonishing skill. One side effect was the virtual disappearance of the decorative qualities that had been so much to the fore in earlier maps. The atlas published by John Tallis was an exception, but it was almost the last of its kind.

The London publisher John Tallis is remembered for two works, *London Street Views* (1838) and the *Illustrated Atlas of the World* (c. 1851), in which this map of Egypt and Arabia Petraea appeared. The quality of its steel engraving is apparent in the detailed mapping, the beautiful decorative border based on ancient Egyptian motifs, and the vignettes of ancient and Arab monuments and scenes. The map and decorations were drawn and engraved by different hands. The map shows Egypt as it was under the dynasty founded by Mehemet Ali, which ruled as nominal vassals of the Turkish sultan. The scientific mapping of the country had begun after Napoleon's Egyptian campaign of 1798, when French engineers and geographers carried out the surveys published as the *Carte Topographique de l'Egypte* in 1808. Tallis's map shows a classical orientation typical of the nineteenth-century British, allocating the Sinai peninsula not to Egypt but to Arabia Petraea, the name of what had once been a Roman province.

SUEZ
Suez, already a port at the head of the Red Sea, though not yet beside a canal. It was supplied with water by the "Fount of Moses," an oasis said to have been miraculously sweetened by the Hebrew leader.

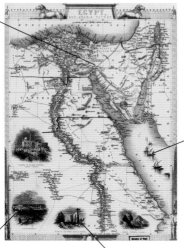

DHOWS
Lateen-rigged dhows on the Red Sea: a tranquil scene that suggests a way of life unchanged for centuries. For all its charm, this image of what was actually a busy commercial seaway is a misleading one.

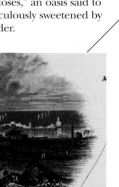

PORT
Alexandria has been Egypt's greatest port since ancient times. It stands on a strip of land that separates the salt Lake Mareotis from the Mediterranean.

ROMANCE
The ruins of the great temple at Karnak, like the other finely engraved vignettes, present a romantic, traveler's-eye view of Egypt. It is depicted as a land of noble, broken antiquities, and picturesque mosques.

THE LURE OF ANCIENT EGYPT

Most of the illustrations on Tallis's map evoke the fallen splendors of ancient Egypt. During the nineteenth century, as France and Britain increasingly dominated Egypt's political history, European scholars and archaeologists played a more disinterested role in understanding, uncovering, and conserving the ancient monuments.

Tallis's map was published at a time when Egypt was effectively independent, owing only a nominal obedience to the sultan of Turkey. Within a few years, however, the country was deep in debt and its government was being run under Anglo-French supervision. After 1881 the French presence disappeared and for forty years Egypt was effectively a British protectorate.

Throughout the period, Europeans were fascinated by the remains of ancient Egypt, which were scattered over the landscape in amazing, monumental abundance. Though never forgotten, Egyptian civilization really gripped the European imagination after Napoleon Bonaparte's famous expedition to the East in 1798–1799. Bonaparte's campaigns achieved nothing from a political point of view, but the association of Napoleon the military genius with the exotic East made both of them fashionable, and the sphinx and lotus became part of the contemporary decorative vocabulary. On the sober scholarly level, the scientists and artists who had accompanied Bonaparte to Egypt took immense pains to measure and document everything they saw, eventually publishing the still awe-inspiring nineteen volumes of the *Description of Egypt* (1809–1822).

The Rosetta stone

The *Description* contained nothing written by the Egyptians themselves—or at any rate nothing comprehensible—since their language was utterly unknown. During the French occupation, however, a soldier found a basalt stone at Rosetta in the Nile Delta with inscriptions in three types of writing. These proved to be hieroglyphics, a less elaborate form of

Above: European fascination with ancient civilizations: craftsmen at work on copies of Egyptian and classical monuments and statuary at London's Crystal Palace in around 1852.

Egyptian writing known as demotic, and an already known language, Greek. If, as seemed likely, all three were making the same statements, detailed comparisons might provide the key to deciphering the language of the ancient Egyptians. By the late 1820s, this formidable task had been accomplished by a French scholar, Jean François Champollion, and the ancient inscriptions and papyri could at last be read.

These, like Egypt's tombs, temples, and monuments, lay at the mercy of treasure- and trophy-seekers who engaged in widespread cultural vandalism for their own or their employers' gratification. Scholars such as John Gardner Wilkinson and Karl Lepsius were indefatigable in recording the remains, but cavalier looting was brought under control only in the 1850s. At that time Auguste Mariette, the first director of the Egyptian Antiquities Service, asserted the state's ownership of antiquities and required all would-be excavators to apply for licenses. Enforcement was patchy at first, but matters improved with the growing influence of serious research bodies like the Egyptian Exploration Fund, established in Britain in 1882.

The most spectacular finds

The age of serious investigation had dawned, with increasing emphasis on artifacts from everyday life rather than on treasures. Nevertheless, there were some spectacular finds: Mariette unearthed the Serapeum and the tombs of sacrificial bulls at Saqqara; Gaston Maspero tracked down a large cache of royal mummies, removed in antiquity from tombs that could not be protected from robbers; and Flinders Petrie, often described as the "father of scientific archaeology," made a host of discoveries including the Amarna letters (Egyptian "Foreign Office" correspondence) and wonderful wax-painted Romano-Egyptian portraits placed in burials.

Though archaeology had become a scientific discipline, there were still discoveries that combined historical significance with the glamor of treasure and adventure. By 1922 all the royal tombs in the Valley of the Kings seemed to have been found. Though very interesting, they had been thoroughly looted in antiquity. A British archaeologist, Howard Carter, was in his last season, having failed to find the unopened tomb he believed to exist in the valley. Then on November 4, as the rubble left by ancient workmen was being cleared away from a spot close to the tomb of Ramses VI, a single step was uncovered. Soon there were sixteen steps leading down to a door. When this was removed and Carter peered through a breach in a second barrier, his patron, Lord Caernarvon, asked whether he could see anything. "Yes, wonderful things," Carter replied, having become the first person for more than 3,000 years to look on the gilded funeral panoply of the young pharaoh Tutankhamen.

Above: The Battle of the Pyramids, July 1798. Here, Napoleon decisively defeated the Egyptian Mamluk army, losing only 40 men in the fighting. Napoleon's brief stay in Egypt failed to achieve its objectives but intensified Europe's fascination with the ancient East.

OVERLAND ROUTE TO INDIA.

POST OFFICE, LONDON

GIBRALTAR

MALTA

NOTE
The Mail Steam Packet Route
The Marseilles Overland Route
The German Overland Route
The Euphrates Route.

THE MAIL CROSSING THE DESERT

SUEZ

ADEN

MADRAS.

The Map Drawn & Engraved by J. Rapkin

OVERLAND ROUTE TO INDIA
JOHN TALLIS

With its charming border and lovely vignettes, this map from Tallis's *Illustrated Atlas of the World*, published about 1851, belies the dynamic reality it records: the network of routes and services that bound together the far-flung British Empire, and especially the home country and British India.

The routes drawn on Tallis's map are not, in fact, overland all the way, but this description implicitly contrasts them with the long passage, entirely by sea, round the Cape of Good Hope. Within twenty years of the publication of Tallis's map, a new sea route came into being with the opening of the Suez Canal, which then became the fastest and most convenient way to reach India and the Far East. The routes marked on the map are easy to follow from the key, even though the inking-over of the lines between London and "Hindoostan" has nullified the color-coding. The very existence of the map bears witness to the significance of India in British minds. During this period the subcontinent was still governed by the East India Company, whose quest for profit had turned it into a political power. By Tallis's day the company was subject to government control and India's governor was appointed by the Crown. Soon, the company would be dissolved and India would become a Crown colony.

KEY TO THE MED
Gibraltar was often described as the key to the Mediterranean, its possession vital to Britain's navy. Its role was enhanced once the main sea route to India passed through the Mediterranean and the Suez Canal.

DESERT MAIL
"The Mail Crossing the Desert." Once the Suez Canal was opened in 1869, this picturesque service was replaced by Royal Mail steamships, which had priority of passage through the canal.

METROPOLIS AND NAVAL BASE
A curiously tranquil view of a great Indian city (present-day Mumbai) that was already a teeming metropolis as well as being the principal naval base in British India.

VITAL PORT
A British possession from 1839, Aden was part of the imperial network of coaling stations. After the Suez Canal was opened in 1869, its position on the sea route to India gave it a vital role.

THE OVERWATER ROUTE

When Tallis's map of overland routes to India appeared, the British were busy building a railroad to connect the Mediterranean and the Red Sea. They scoffed at the French notion of constructing a canal, but—as much by luck as good judgment—became the nation that profited most by it.

I n modern times the first serious proposal to build a canal was made during the brief French occupation of Egypt led by Napoleon Bonaparte in 1798–1799. Preliminary reports were discouraging and the matter was taken no further at the time, but French interest persisted. In the 1830s the Egyptian pasha, Mehemet Ali, was being urged to back a canal by French interests, while the British agitated for a railroad from Alexandria on the Mediterranean, via Cairo, to Suez on the Red Sea. Apart from the enticing prospect of lucrative railroad contracts, British governments viewed the canal as likely to be a strategic disaster. While the route round the Cape of Good Hope remained the principal path to the East for ships, people, and munitions, Britain's position in the Atlantic gave it a distinct advantage. However, the opening of a Suez canal would, it was argued, transfer that advantage to the navies of France and other Mediterranean powers. In 1847, when a three-man international commission reported that a canal was not feasible, the French suspected, with some justification, sabotage by the British member, railroad engineer George Stephenson.

The tables turned

The shrewd Mehemet Ali was happy to avoid committing himself to a railroad or a canal, realizing that either would lead to interference in Egyptian affairs by European powers. Mehemet's successor, Abbas, gave way to British pressure and in 1851 signed a contract for the building of a railroad. In 1854, however, while the railroad was still under construction, Abbas was assassinated and the tables were turned. The new pasha, Said, invited a friend from his boyhood to Egypt, former French diplomat Ferdinand de Lesseps.

Above: Ferdinand de Lesseps became associated with the Suez Canal thanks to a childhood friendship; later, and disastrously, he championed a scheme for a Panama canal.

Though neither an engineer nor a financier, de Lesseps was obsessed with the idea of building a canal, and his enthusiasm rapidly converted Said. In November 1854 the syndicate that de Lesseps proposed to form was granted a ninety-nine-year concession and exemption from paying taxes, and was promised the services of virtually unlimited Egyptian labor. Said contented himself with a mere 15 percent of the profits and, when de Lesseps experienced difficulties in raising the capital, himself purchased 44 percent of the shares.

The British completed their railroad and did all that they could to hinder de Lesseps. However, on April 25, 1859, work began, and over the next ten years a colossal excavation drove the canal in a straight line from what would become Port Said, through a series of lakes, and on for a farther 15 miles (25 km) to the Gulf of Suez.

The work was not seriously interrupted when Said's successor, Ismail, managed—for an exorbitant price—

Above: A fleet of ships enters the northern end of the Suez Canal at its official inauguration, on November 17, 1869. The event was stage managed by the Egyptians to create a spectacle that was heard of around the world.

to reduce the burden on Egyptian labor. A reckless spendthrift (though often on good causes such as schools), Ismail turned the opening of the canal in 1869 into an extraordinary spectacle, attended by a gathering of European royal personages, ambassadors, and literary figures. But contrary to legend, this was not the occasion on which Verdi's opera *Aida* was performed for the first time.

The British change their minds

Still hostile, the British government was not represented at the celebrations. Now supposedly convinced that the new route would not make any great difference to the shipping lanes, the British soon discovered that in fact three-fifths of the traffic through the canal was flying their own flag. Moreover, British notions of the canal's strategic value had to be revised as the perception of a Mediterranean "threat" to India was replaced by fear of a Russian advance from Central Asia. In 1875, when a desperate Ismail decided to sell his shares, the British prime minister,

Benjamin Disraeli, stepped in and acquired them for just under four million pounds. Within a short time the British had become convinced that Suez was a vital link in imperial communications. As the largest single shareholder in the canal, Britain began a momentous involvement in Egyptian affairs, and a military presence in the Canal Zone, that would end only in 1956.

Above: The Suez Canal is today one of the world's busiest shipping lanes, used by more than 50 vessels every day.

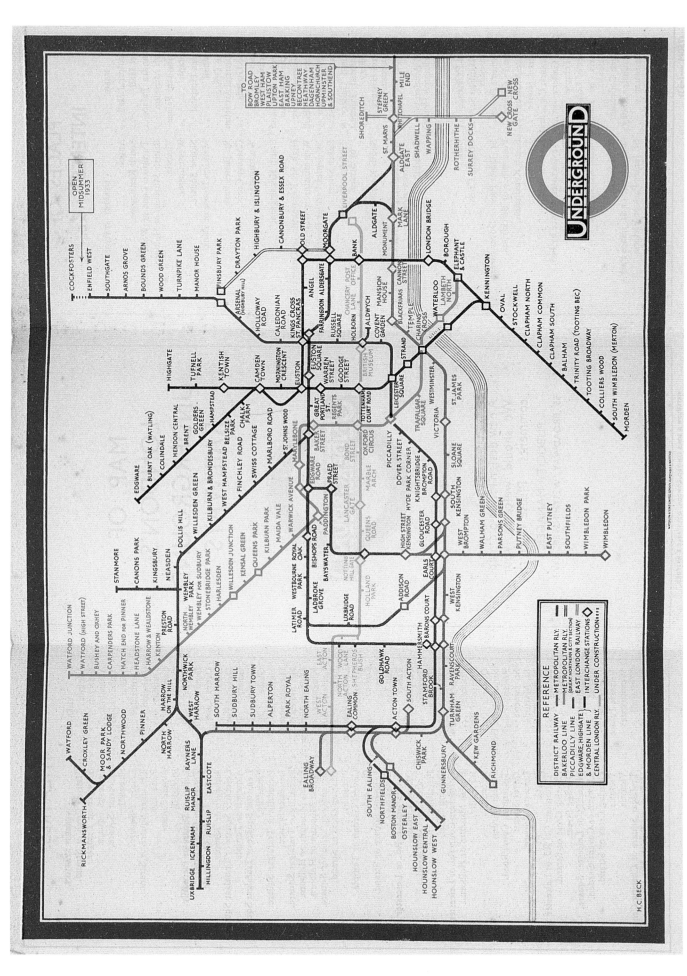

LONDON UNDERGROUND
HENRY C. BECK

Harry Beck's map of the London Underground is now a classic of twentieth-century design—visually appealing, simple to understand, and thoroughly functional. At the time of its first appearance in 1933, it was regarded as revolutionary, but it soon became the model for underground transport maps everywhere.

Between the world wars, London Underground played a leading role in contemporary design, pioneering a bold new typographic style and eye-catching posters inspired by modern art movements such as cubism. Beck was employed by London Underground as a graphic draftsman. He developed a personal interest in making a simpler, more intelligible map of the system. The existing map was topographically accurate, but the crisscrossing of lines was confusing, it was hard to sort out the most efficient route between stations on different lines, and the layout was unpleasing to the eye. Beck's solution was to abandon geographical correctness. His map, possibly inspired by electric circuit diagrams, was essentially geometric, with mostly straight lines and angular relationships, color coding, and stations indicated at intervals that left space for legible labeling. Designed in 1931, Beck's map was initially rejected. A subsequent trial proved that the public liked it, and in January 1933 it was officially issued.

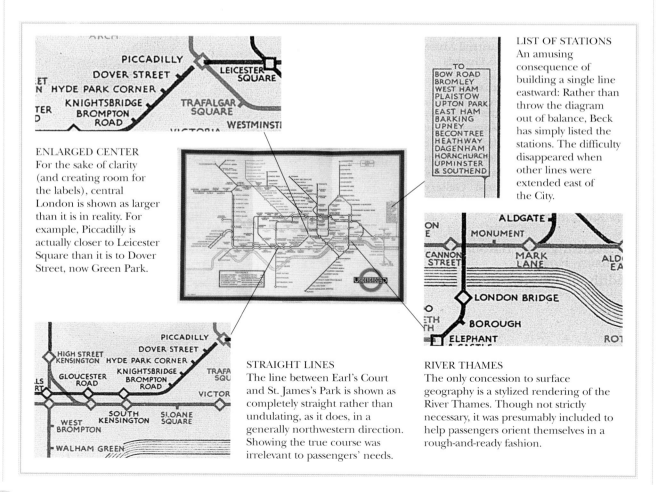

ENLARGED CENTER
For the sake of clarity (and creating room for the labels), central London is shown as larger than it is in reality. For example, Piccadilly is actually closer to Leicester Square than it is to Dover Street, now Green Park.

LIST OF STATIONS
An amusing consequence of building a single line eastward: Rather than throw the diagram out of balance, Beck has simply listed the stations. The difficulty disappeared when other lines were extended east of the City.

STRAIGHT LINES
The line between Earl's Court and St. James's Park is shown as completely straight rather than undulating, as it does, in a generally northwestern direction. Showing the true course was irrelevant to passengers' needs.

RIVER THAMES
The only concession to surface geography is a stylized rendering of the River Thames. Though not strictly necessary, it was presumably included to help passengers orient themselves in a rough-and-ready fashion.

PIONEERING UNDER GROUND

By the mid-nineteenth century, great industrial and commercial cities were becoming congested as travelers crowded into the streets on the way to work or in pursuit of pleasure. The remedy—a city-wide system of underground passenger trains—was first attempted in Victorian London.

Above: The Underground was a response to traffic congestion, but today the average speed on London's roads remains 8 mph. (12 kph), virtually the same as in the horse-drawn Victorian era.

When Britain's overground railroads reached London, they terminated around the edges of the metropolis. Most travelers were then compelled to take cabs, horsedrawn omnibuses, or trams to their innercity destinations, passing through traffic-jammed streets. There were proposals for a central station to which all the existing lines would be connected, but the expense and difficulty of acquiring and building on densely occupied land were prohibitive. In 1853 Parliament authorized an underground line linking two north London termini, but there was little public enthusiasm for what was seen as an unnatural mode of travel, and the scheme stalled for years.

Finally, with contributions from the Corporation of the City of London and the Great Western Railway, work began in 1859 on what became the Metropolitan

Above: Dignitaries, including later prime minister William Ewart Gladstone (fourth right), join directors of the Metropolitan Railway Company to inspect the world's first underground line in May 1862. The line opened the following year.

Above: The first train on the Metropolitan Line passes beneath Praed Street in 1863. Like overground trains, engines on "the Sewer" were powered by steam, making the experience of traveling hot and noisy.

line. The method of excavation, known as "cut-and-cover," involved digging up the street, constructing the (relatively shallow) tunnel, and then covering it up again. In June 1863 service began between Paddington (terminus of the Great Western Railway) and the City (Farringdon Street), via King's Cross (terminus of the Great Northern Railway) and other stations. On the first day some 30,000 passengers, mostly sightseers, traveled underground for 3½ miles (5½ km) in what was irreverently known as "the Sewer."

The underground system grows

Despite early confusions—a variety of train types and two different gauges—the Metropolitan line was a success. The company planned extensions, and Parliament considered schemes to create a circular underground system for central London. A second company, the Metropolitan District, was established to build just north of the Thames, and at the end of 1868 its line was linked with a greatly expanded Metropolitan line to form a near-complete circle. Mainly thanks to rivalries between the companies, however, the last five stations, through the City of London from Mansion House to Aldgate, were not opened until 1884, creating what was in effect the present-day Circle line.

By that time the system was already spreading east and west. The trains currently in service were (like all overground trains) driven by steam traction, and the tunnels were still being constructed by cut-and-cover. The next step forward was the projection of the City of London and Southwark Subway, intended to relieve the severe traffic congestion on the routes from south London into the City. This involved tunneling beneath the Thames, which disqualified both cut-and-cover and steam.

Sir Marc Brunel had driven a tunnel beneath the Thames as early as 1843, but at an enormous cost in terms of money and labor. However, in 1880 the civil engineer James Henry Greathead had developed a new "shield" system of boring circular tunnels through soft soil, lining them with cast-iron rings, and protecting them externally with concrete. A single circular bore was used for each of the two tracks, and—stable London clay proving the ideal substance for the Greathead shield—the project was carried through faster and much more cheaply than on the cut-and-cover lines. The "tube" form of tunnel gave rise to the name used by Londoners for the entire network, "the Tube."

The first electrified railroad

When the new line was originally envisaged in 1884, it was intended that trains should be moved by the unattractively slow method of cable traction. By 1888, though, the subway company had been converted to the idea of using electricity, and in 1890, on the opening of the original 1¼-mile (2-km) section from the City (King William Street) to Stockwell, the City and South London line became both the first deep-level and the first electrified railroad in the world. Though the line was in many ways unsatisfactory—it turned out to have been conceived on far too small a scale—it became the pattern for the long future awaiting the London Underground.

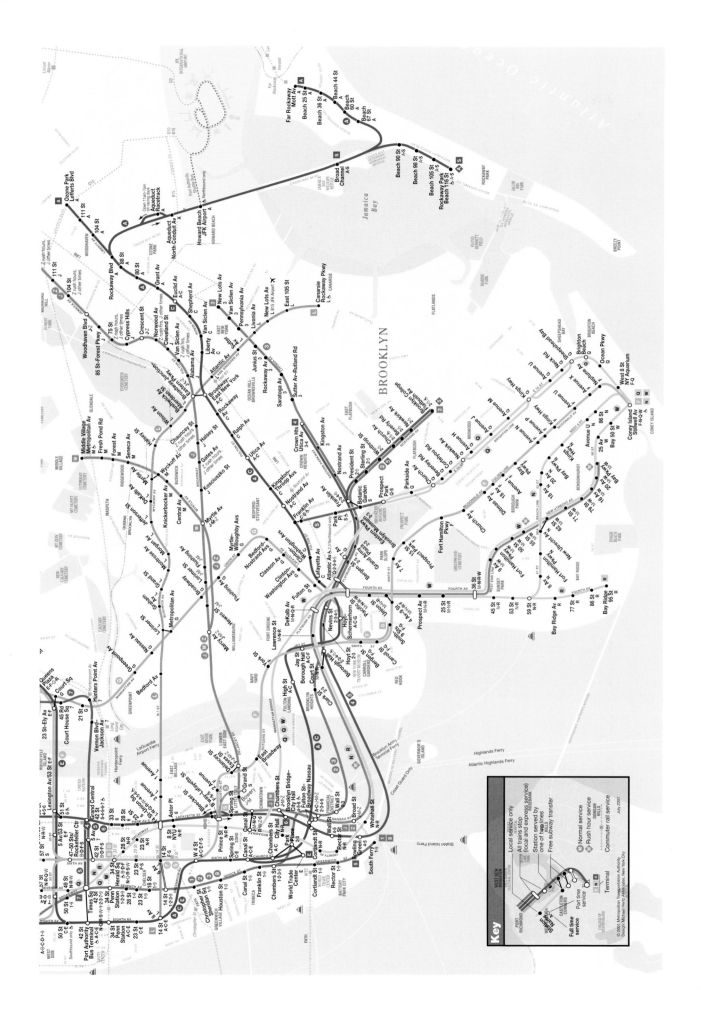

NEW YORK CITY SUBWAY
MAPPING UNDERGROUND

The most extensive underground system in the world, the New York City Subway was constructed by engineers who learned from the earlier achievements and mistakes of London and Paris. The map incorporates a wide range of information, using a variety of colors and symbols to achieve clarity.

Faced with congestion problems similar to those of London, New York did not at first adopt a system based on underground travel. Instead, from 1870 an elevated railroad, the "El," was constructed to carry passengers in trains that ran through overhead galleries above the streets. By the 1890s possible extensions of the light transit system were being discussed in terms of subways. New York's first subway project was an example of large-scale planning: The 9.1-mile (14.6-km) system was laid out in a single three-and-a-half-year operation and opened in October 1904. Intelligently anticipating demand, four rails (two parallel tracks) were installed to accommodate both local and express trains. Like the El, the subway took advantage of the city's wide, straight streets, its tunnels being positioned to avoid housing. A type of cut-and-cover construction was used (digging deep trenches on one or both sides of the street and then excavating horizontally) that minimized interference with road traffic.

BUSY STATIONS
The busiest stations on the subway are Times Square, Grand Central, and 34th Street-Herald Square. In 1999 a combined total of 110 million passengers passed through the stations' turnstiles.

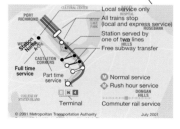

CITY HALL
For obvious reasons, City Hall was built as the "jewel" of the first subway line. It was opened on October 27, 1904, when Mayor John McLellan drove the first train to 103rd Street.

KEY
The five colored bands are visual aids, not lines. The twenty-five routes are indicated by numbers or letters, each in a colored box whose shape is determined by its function at a particular location.

TO THE RESCUE
The Franklin Avenue Shuttle, completely reconstructed in 1999, is one of a number of initiatives in the 1980s and 1990s that rescued a system that had apparently gone into an irreversible decline.

SUBWAY STORY

New York City's subway system has experienced extremes of prosperity and decay. Both agent and victim of social change, it has also been affected by the often tangled finances of its authorities, among them a not-always-solvent NYC. Apparently doomed in the 1970s, the subway was rescued by a determined effort that restored its status as a world leader.

The subway might have been built as much as thirty years before its opening in 1904 if pioneering schemes had not been blocked by vested interests. The most powerful of these was headed by "Boss" William Marcy Tweed, whose corrupt political machine ran New York. Thriving on rakeoffs from established services, Tweed opposed any new development that he could not control.

One ingenious inventor and entrepreneur, Alfred Ely Beach, did manage to fool City Hall. Under cover of a license to construct a pneumatic mail-dispatch system, he built a block-long subway between Warren Street and Murray Street. Designed on a novel

pneumatic principle (cars were driven down a tube by giant fans), it was opened to public acclaim. Even so, Tweed was able to prevent any extension of the line until 1872–1873, when he was charged and convicted for his thefts—on 204 counts. But it was too late for Beach's project, for a Wall Street crash had frightened off potential investors. The pneumatic subway was never revived, and New York had to wait thirty years for one based on a more conventional technology.

In 1904, when the Interborough Rapid Transit (IRT) system came into operation, its twenty-eight-station Manhattan route ran from City Hall to 145th Street and Broadway. The service was an instant success, and proposals for expansion were soon being tabled. After much lobbying, contracts were signed with two companies, the IRT and the Brooklyn Rapid Transit Company (BRT). Between them they extended the subway to three more of the city's boroughs: the Bronx, Brooklyn, and Queens.

For a statutory five-cent ticket, passengers could travel anywhere on the network. Over the years this became an excellent, and finally superb, value. Cheap citywide travel made New Yorkers extremely mobile,

Above: N.Y.P.D. officers stand by as a Broadway Local subway train stops to take on passengers in a photograph from the early 1900s.

Above: Graffiti were the least of the subway's problems in the early 1980s, when this picture was taken. Falling investment, regular breakdowns, and frequent muggings had reduced passenger numbers.

and when groups that were prospering abandoned their old neighborhoods, they tended to move outward along the subway routes.

For the subway companies, which had originally insisted on the nickel fare, making a profit became more difficult as other prices rose. The BRT actually went bankrupt in 1918 and was reorganized as the Brooklyn–Manhattan Transit Corporation (BMT). The City of New York also suffered, since the contracts with the subway companies stipulated that they could take their guaranteed profits before paying for their leases from remaining funds, if any. In practice, the amount of revenue the city received from the subway was paltry.

Proposals for the municipal control of new lines therefore had appeal, and in 1925 work began on a city-owned Independent Subway (IND). The first section was opened in 1932, by which time the IRT was also in deep financial trouble. In 1940 the city bought the IRT and BMT, creating a unified municipal system.

After flourishing during World War II, the subway began a long, slow decline in the face of falling passenger numbers and booming automobile use. In 1947 the nickel fare was finally raised to ten cents,

but neither further increases nor the creation of a separate New York City Transit Authority in 1953 significantly improved the situation.

The subway in decline

Even so, new lines and connections continued to be built until the 1970s. Then in 1975, with the city itself nearly bankrupt, construction on the new Second Avenue subway ceased and all the holes that had been dug were simply filled in. Investment in the decaying infrastructure actually fell, and the subway appeared to be in terminal decline. By this time it was notorious for dirt and litter, graffiti-covered cars, breakdowns and cancellations, and the frequent crimes committed in stations or on trains.

An astonishing transformation began in 1982, when a program was launched to refurbish, modernize, and improve security on the system. Energy and investment paid off spectacularly, a success symbolized in 1989 by the withdrawal of the last graffiti-covered train. The new policy was sustained during the 1990s, bringing the passengers back and taking the system into the twenty-first century in better shape than ever.

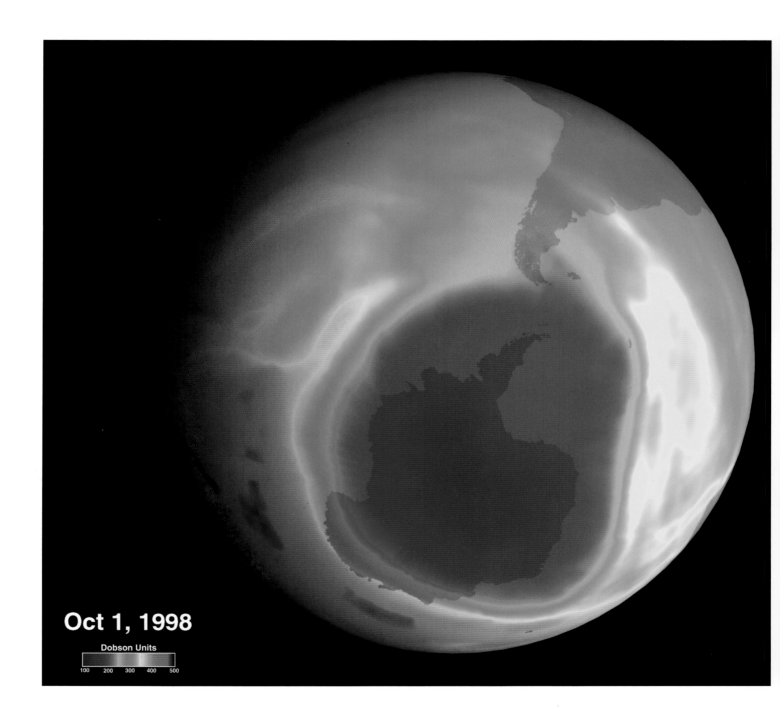

Oct 1, 1998

Dobson Units

100   200   300   400   500

THE OZONE LAYER
FALSE-COLOR SPECTROGRAPH, 1998

The ozone layer protects the Earth from harmful solar radiation. Few people were aware of this until the 1980s, when the depletion of the layer, and its consequences, were widely publicized. A prompt international response appears to have averted the worst, but it is predicted that the ozone layer will not return to normal before the late twenty-first century.

Ozone exists in the upper atmosphere in small but vital amounts. It absorbs ultraviolet solar radiation, which would otherwise have a devastating impact on organic life. In 1985 Joe Farman, a scientist working for the British Antarctic Survey, reported a serious thinning of the ozone layer over Antarctica. Skepticism faded as the thinning grew worse, and previously neglected data revealed that the depletion had been going on since the 1970s. The main cause was identified as chlorofluorocarbons (CFCs), industrial chemicals used in refrigerators, air-conditioning, aerosols, and foam packing. They broke up in the atmosphere and released chlorine, which in turn reacted with ozone, converting it to oxygen. Among the effects of increased solar radiation were skin cancers, eye disorders, immune system damage, and the destruction of phyloplanktons crucial to the ocean food chain. By 1990 the "ozone hole" was a dozen times larger, and a series of international agreements arranged for the phasing out of CFCs.

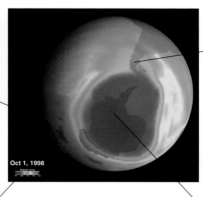

DANGER ZONE
Tierra del Fuego, the Falkland Islands, and the southern tip of mainland South America are clearly in the danger zone, and threatened with worse if the "hole" should drift or become enlarged.

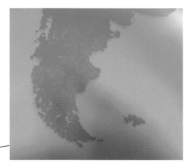

AUSTRALASIA
New Zealand was clearly out of danger in October 1998. There is evidence, though, that the hole can drift on occasion, lowering levels above both New Zealand and Australia.

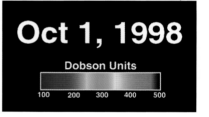

DOBSON UNITS
October is springtime in the Southern Hemisphere. Dobson units are a measure of the proportion of ozone in the upper atmosphere; one unit represents about 4 molecules of ozone per square inch (27/sq. cm).

ANTARCTICA
At the time when this map was made, Antarctica lay almost directly under the thinnest part of the ozone layer, represented by the purple-blue end of the color range.

GLOBAL WARMING

The depletion of the ozone layer has been an unmistakable symptom of the damage done to the environment by human activity. Even if the ozone layer is repaired, there is still the prospect of a rapid and destructive change in the climate as the Earth becomes progressively warmer.

Weather-borne catastrophes feature in the news with increasing frequency. Hurricanes, floods, droughts, and storms devastate communities, take lives, and inflict huge financial losses on householders, businesses, and insurers. Heat waves and other kinds of extreme weather have also become more common.

The prime cause has been identified as a rise in global temperatures that is becoming steadily more pronounced. Data from a variety of sources confirm that global warming is taking place, and its rapidity is illustrated by the fact that ten of the eleven warmest years on record have occurred since 1980.

However, if the phenomenon is undeniable, the reasons for global warming have been much disputed. As evidence from ice cores, tree rings, and other natural "time capsules" has demonstrated, abrupt climatic changes have taken place in the past, long before human activity could have had a worldwide influence. So global warming could be, partly or entirely, a natural process.

However, most scientists now emphasize the role of human agencies. Like the panes in a greenhouse, gases released into the atmosphere trap the solar rays reflected off the Earth's surface, raising its temperature. Among these "greenhouse gases" are carbon dioxide, methane, nitrous oxide, ozone, and CFCs. The CFCs are being phased out, and it is predicted that their impact will begin to decline. However, the other gases are the result of processes closely bound up with the functioning of advanced economies, tempting the world's leaders to deny the reality of global warming or at least defer the necessary measures.

This is particularly the case with carbon dioxide. The gas is produced by most living things, which breathe in oxygen and breathe out carbon dioxide; life is able to exist on the planet because the process is reversed in plants and trees. The carbon dioxide in the atmosphere is increased when fossil fuels such as wood, coal, and oil are burned and when trees are cut down—two activities that have occurred on a global scale in recent times.

The burning of fossil fuels intensified from about 1800 with the industrial revolution, and in the twentieth century the effects of economic expansion far outweighed any increases in efficient energy use. Though not new, deforestation became a cause for anxiety in the late twentieth century, when vast areas

Above: Low-lying islands—this is Ihuru in the Maldives in the Indian Ocean—are one of the most vulnerable habitats to the rising sea levels that will result if global warming continues.

Above: A haze of smog, caused by the interaction of carbon dioxide and sea air, obscures downtown Los Angeles.

in the tropics were cleared for agricultural use or timber sales. As a result of these developments, the amount of carbon dioxide in the atmosphere rose from about 280 parts per million (ppm) in 1800 to 360 ppm in 1995.

The effects of global warming

Among the known effects of global warming are rising sea levels, which threaten low-lying islands; some of these, such as the Maldives, will disappear from the map if present trends continue. Famous cities such as Venice are under threat, and many countries will suffer erosion and inundation unless they install expensive flood barriers. Other regions face the opposite problem: an increased dryness that brings summer forest or bush fires to places as remote from each other as the Mediterranean and Australia. Warming also intensifies desertification and crop failures, causing widespread starvation in parts of Africa.

Other effects include changing seasonal times, the ability to grow certain plants farther north, and the northward march of mosquitoes and locusts, epidemics, and crop damage. Much freak weather has been attributed to global warming, and many of its effects are, as yet, unforeseeable. One plausible theory is that melting polar ice could disrupt the Gulf Stream that keeps northwest Europe relatively warm, so that temperatures in the region could actually plunge.

The most obvious remedy—using less fossil fuel—is the most difficult for economies in which industry creates prosperity and automobiles are part of everyone's way of life. The development of first-class public transport services and nonfossil-burning energy sources evidently requires heavy expense and great political will; recycling and tree-planting have proved somewhat easier to popularize. At the beginning of the twenty-first century, the world's leaders are still struggling to come to terms with the problem.

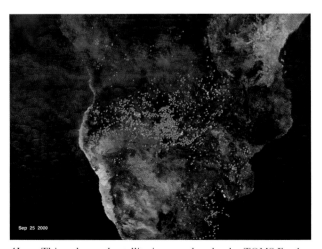

Above: This enhanced satellite image taken by the TOMS Earth Probe shows wild- and land-clearance fires raging across southern Africa in September 2000.

Hurricane Hugo

September 21, 1989 18:44 GMT
NASA-GSFC Lab for Atmospheres
Hasler, Pierce, Starr, Jentoft-Nilsen

Derived from NOAA AVHRR
RGB = 0.65 μm, 0.9 μm, 11μm

HURRICANE HUGO
NASA GODDARD LABORATORY

An image of overwhelming power, this satellite weather map was produced by the NASA Goddard Laboratory for Atmosphere. It shows Hurricane Hugo relentlessly on course to strike the state of South Carolina and wreak more than $7 billion in damages, making it one of the most expensive hurricanes ever recorded.

L ate in September 1989, Hurricane Hugo devastated Puerto Rico and then struck the southeastern United States. It tore a trail of damage 150 miles wide through South and North Carolina, wreaking terrible damage on the cities of Charleston and Myrtle Beach. There were mass evacuations as houses were flattened and boats, cars, and planes piled into splintered heaps of wreckage by winds that reached 150 miles (245 kilometers) an hour. After Hugo struck Puerto Rico, where the destruction triggered riots and looting, meteorologists speculated that it might swing north and miss the continental United States. In fact, it regained strength as it headed straight for South Carolina, underlining the limits of meteorology as a predictive science. The huge energy they unleash makes hurricanes irresistible, and even though satellites, planes, and radar can track them and warn threatened communities, their notorious unpredictability adds to the fear and disruption they cause.

VICTIM
Florida escaped Hurricane Hugo but in 1992 it was the victim of the United States' most expensive hurricane, when Andrew devastated both Florida and Louisiana. The total bill exceeded $20 billion.

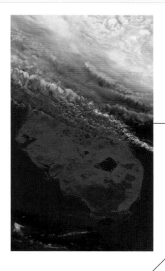

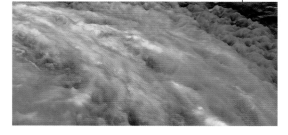

EYE
The strangely calm center around which the hurricane rages. For once a well-worn metaphor— "the eye of the storm," signifying a peaceful moment in a turbulent situation—is irreproachably accurate.

SWIRLING
In this area of the hurricane its structure is particularly clear, with great columns of swirling cloud, looking like moving mountains of ice, feeding into the vortex.

FORECASTING THE WEATHER

Hurricanes, tornadoes, and the like are extreme examples of forces that affect human beings everywhere. If the weather cannot be controlled, we can at least adapt to it—provided we know what to expect. As a consequence, charting and forecasting the weather has become a global scientific endeavor.

From the earliest times, the weather was of vital concern to hunters, farmers, warriors, and sailors. More recently the list has extended to industry, aviation, transport networks, and tourists. Meanwhile, governments and their agencies have had to make plans to cope with the consequences of rising global temperatures and increasingly unpredictable natural events.

Theories about the behavior of the elements can be traced back to the Babylonians, and most cultures have developed folk wisdom of variable accuracy on the subject. Important advances took place in the seventeenth and eighteenth centuries, when essential scientific concepts and instruments were devised. However, although some individuals kept records of weather conditions, systematic and sustained observation began only in the nineteenth century; the first in the field, in 1814, was the Radcliffe Observatory at Oxford, England. In the United States and Europe, national weather bureaus were established in the course of the century, and the first weather maps are said to have been on sale at the 1851 Great Exhibition in London. Cooperation across national boundaries began in 1873 with the founding of the International Meteorological Organization, now the World Meteorological Organization (WMO).

As in other disciplines, scientific and technological progress accelerated after World War II. The global reach of observation and forecasting became more pronounced as the complex and worldwide nature of weather systems was realized. Today, thousands of weather stations receive data every six hours from observers who take readings of temperatures, rainfall, humidity, wind speed and direction, air pressure, and the type, height, movement, and extent of clouds.

Information also arrives from ships, aircraft, and oil rigs; from radar and electronic instruments on buoys and balloons; and from stationary or orbiting satellites whose sensors can measure a range of phenomena, such as ground temperatures and wind speeds. The vast quantities of data are fed into computers capable of making billions of calculations. They create global computer models of current conditions, and also manipulate the information to simulate the future behavior of the weather.

Synoptic and prognostic charts

The computers' findings become the basis of weather maps. Those describing existing conditions are known as synoptic charts; prognostic charts are forecasts of likely developments. A large number of internationally recognized symbols provides the required detail—for example, distinguishing between six states of snow and rain, including "intermittent, moderate," "steady,

Above: Oxford's Radcliffe Observatory pioneered systematic observation of the weather; in the United States the Smithsonian was producing daily weather maps by 1849.

Above: Automated weather stations, like this one at Halley Bay in Antarctica, constantly monitor local weather conditions.

moderate," and "intermittent, heavy" falls. Among the most immediately recognizable features of weather maps are the isobars—lines that link areas of equal air pressure (just as a contour line on a physical map links areas of equal height). Areas of low or high pressure are marked, but are in any case identifiable according to whether the isobars are closely packed (low pressure, generally signifying strong winds) or more widely separated (high pressure, generally signifying calm conditions).

Also prominent are the fronts, representing the boundaries between masses of warm and cold air, which usually portend a change in the weather. A dark line studded with triangles signifies a cold front, whereas a warm front is depicted by a similar line with dark semicircles. Temperatures, wind speeds, and shaded areas indicating rainfall are among the many other standard features on weather maps, which are reproduced in simplified form in newspapers and on television.

The huge capacity of modern computers, and an improved understanding of weather systems, have made forecasting far more reliable—short-range predictions (24–48 hours) are now said to be 85 percent accurate. Lives are saved by specific applications such as the Doppler Radar System, which can detect changes within storms and make possible faster warning times. There are still failures, however—for example, in October 1987 devastating storms struck Britain shortly after a TV weatherman had assured viewers that no such thing was about to happen.

In any case, the inescapable limitations of forecasting have been increasingly accepted since the development of chaos theory, which recognizes the existence of ultrasensitive systems that can be affected disproportionately by tiny events, so that their behavior is not predictable even in principle. At times this is evidently true of globally linked weather patterns, so that models and maps remain imperfect, though valuable, representations of reality.

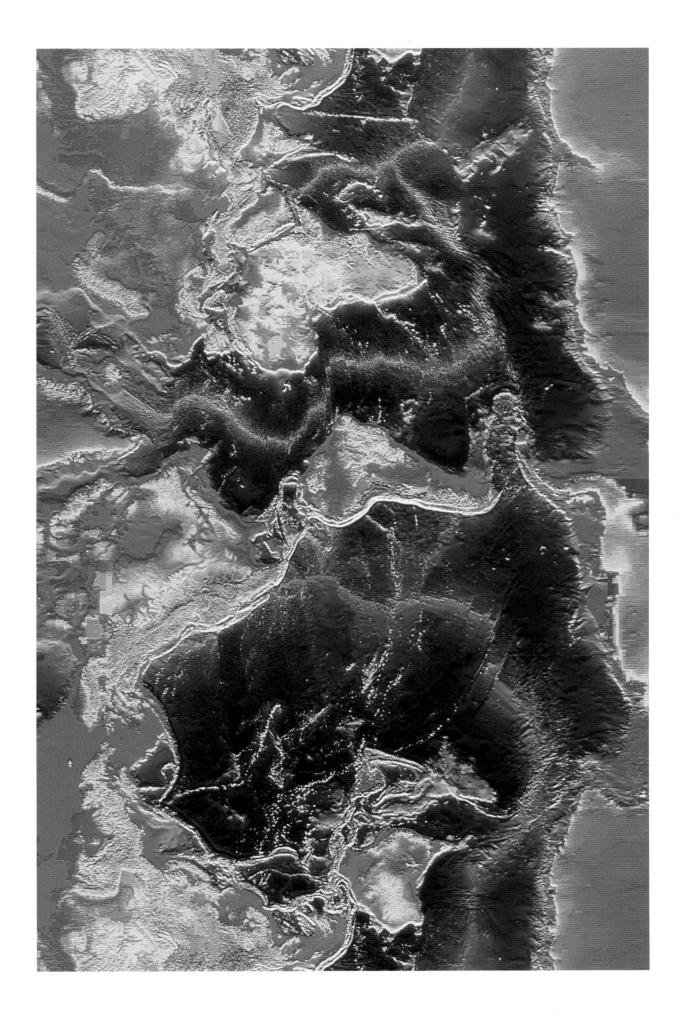

THE OCEAN FLOOR
THE UNDERSEA WORLD

Covered in water that is 2 to 3 miles (3 to 5 km) deep, the ocean floor has long been the least-known region of the Earth. Today, efforts to explore it have been enhanced by maps based on marine and satellite technology. These have revealed the topography of the seabed and clarified the processes that create it.

This 1997 map was made by using results from two methods of investigation. One was data obtained by sonar, directing sound waves at the seabed so that the echoes can be translated into computer-enhanced images. Though informative, the technique covers relatively small areas and is very expensive. Here it has been combined with the wider-ranging findings of satellite altimetry, the measurement by radar of a satellite's height above sea and land surfaces. Because of variations in the Earth's gravitational field, there are bulges and dips in the surfaces of the oceans that relate to the contours of the seabed below. By comparison with landmasses, the ocean floor is geologically simple. Its changing configuration is created by plate tectonics, the movements of the great blocks making up the Earth's crust. Plates may grind past each other, move apart to form new floor, or collide. "Old" plate is continually disappearing as "new" is formed, and so the ocean floor is, in geological terms, ever youthful.

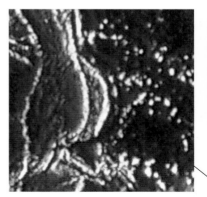

HOT SPOT
The Hawaiian chain, each island of which was created by the northwest movement of the Pacific plate over a "hot spot" where magma breaks through to form a volcano. As each island moves past the hot spot, its volcanic activity diminishes.

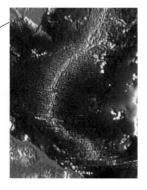

DEEP TRENCH
The Marianas Trench (center), the Earth's deepest depression, is over 7 miles (11 km) deep in places. It is part of the line of crustal trenches, visible round the north Pacific, directly related to the volcanic "ring of fire."

FRACTURE ZONES
Breaks in the ocean floor create series of fracture zones or slip faults; these often cross undersea ridges, creating sheer drops and giving the ridge a zigzag outline.

MOUNTAIN RIDGE
The Mid-Atlantic Ridge (center) bisects the ocean. It is one of a number of long undersea mountain chains where magma forces its way through the crest, creating new undersea crust.

EXPLORING THE DEPTHS

People sailed and fished the oceans for millennia, but they had no way of finding out what lay at any distance beneath them. Scientific studies brought considerable knowledge, but direct experience of the depths was achieved only in the twentieth century.

Ancient Greek thinkers tried to classify the denizens of the seas, and legend has it that Alexander the Great tried to study them by having himself lowered into the sea in a glass diving bell. Use of this device was based on the observation that a bell-shaped object could be lowered into water, trapping breathable air within it. Early experiments with submersibles are more likely to have been made with animal skins, and this was still the case when eighteenth-century English inventors took up the idea, adding a pump to renew the air supply.

The promise of greater mobility made individual deep-sea diving suits more attractive, especially for salvage operations. The basic design, with a helmet attached to a watertight suit and a continuous air flow, dates back to 1819, although almost a century elapsed before problems of decompression—adjusting the diver's body to the varying water pressure—were understood and overcome, allowing divers to work several hundred feet down in safety.

Meanwhile, oceanography was becoming a science. An American naval officer, Matthew Fontaine Maury, compiled and classified the available data in his pioneering *Physical Geography of the Sea* (1855). Then the first systematic on-the-spot study of the oceans was undertaken by a British vessel, the *Challenger*. As *Challenger* circled the globe in 1872–1875, a group of scientists led by a Canadian, John Murray, took soundings and used dredges to collect samples from all the ocean beds. They investigated and measured the Mid-Atlantic Ridge, identified 4,700 previously unknown marine species, and verified the existence of life at unexpected depths, down to almost 27,000 ft. (8,200 m), which they recorded in the Marianas Trench. Their discoveries were eventually published in no fewer than 50 volumes that effectively laid the foundations for the science of oceanography.

None of this actually involved a descent into the depths. That was achieved when an American, William Beebe, invented the bathysphere, a pressure-resistant steel globe that was lowered into the water by a cable.

Right: Pioneer undersea explorer Auguste Piccard, in the light-colored life jacket, emerges from the bathyscaphe *Trieste* in the Mediterranean in October 1953 after making what was then a record dive to 10,355 ft. (3,155 m) beneath the ocean.

Above: The prow of the doomed ocean liner *Titanic*, photographed by an unmanned submersible on the floor of the Atlantic Ocean at a depth of around 13,100 ft. (4,000 m). The wreck was discovered by U.S. oceanographer Robert Ballard in 1985.

It was launched in 1930, carrying Beebe and Otis Barton. A number of descents followed, and in 1949 Barton reached a depth of 4,500 ft. (1,350 m).

The invention of the aqualung
Individual exploration, admittedly at much lesser depths, became possible after 1943, when Jacques Cousteau and Emile Gagnan invented a portable underwater breathing apparatus, the aqualung. Cousteau, a French naval officer, developed waterproof cameras and lights, and produced more than 80 documentary films. The book and film of *The Silent World* (1953 and 1956) made him famous. His work was profoundly influential in promoting an understanding of the beauty and variety of the underwater world, and in making possible direct experience of it through scuba diving.

The limitations and dangers of its cable attachment restricted the usefulness of the bathysphere. From the late 1940s, a Swiss scientist, Auguste Piccard, was developing the unattached, free-moving bathyscaphe, which was driven by an electrically powered motor

and returned to the surface by shedding iron ballast held in place by electromagnets. The first successful descent was made in 1953, and in 1960 the *Trieste*, an improved version acquired by the U.S. Navy, was taken down to the floor of the Marianas Trench (35,800 ft.; 10,900 m) by Lieutenant Don Walsh and Auguste Piccard's son Jacques.

After this, submersibles developed rapidly. Though the term "submersible" describes any underwater vessel, it usually refers to small craft, holding up to three people or remote controlled, which can penetrate and move about in the ocean depths. Equipment, such as arms and grabs, allows operators to take samples and perform other tasks on the seabed.

Submersibles have enabled scientists to investigate intriguing phenomena such as thermal vents, and they are used for checking the condition of oil platforms and pipelines. They also support such high-profile activities as those of Robert Ballard, an American oceanographer associated with the discovery and investigation of famous wrecks, ancient and modern, including the *Bismarck* and the *Titanic*.

EL NIÑO
PACIFIC WEATHER PATTERNS

This satellite image of the Pacific in 1997 uses false colors to dramatize an El Niño event: the arrival of an unusually thick layer of warm water in the eastern region of the ocean. This apparently localized occurrence is now understood to have global consequences, such as drought in Indonesia and Australia and high rainfall along the east coast of South America.

In the usual Pacific weather pattern, trade winds drive warm surface water west, allowing upwelling cold water to replace it. Every few years, however, an "El Niño event" takes place: The winds die down or reverse, and warm currents confine the cold water to deeper levels. El Niño was first observed in 1972 by fishermen in Peru and Ecuador, who suffered from the failure of the nutrient-rich cold water to rise and sustain the anchovy and mackerel harvest. The 1972 El Niño had a ruinous impact on what had been a major industry, subsequently repeated during the event's periodic recurrences. El Niño is in fact one aspect of the complicated, fluctuating relationship between the equatorial ocean and the atmosphere. It has become known as "ENSO," the El Niño Southern Oscillation, although the term omits ENSO's other extreme, La Niña, an unusually strong cooling phase. The cycle's irregularity makes it difficult to anticipate. The greater frequency of recent El Niños may be related to volcanic activity or to global warming.

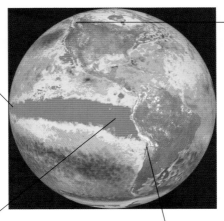

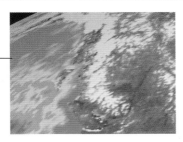

HEIGHT AND WARMTH
This satellite image shows the height as well as the warmth of sea surfaces; hence the irregular appearance of the hemisphere, apparent in areas such as the one shown here.

MILDER WINTERS
El Niño affects nonequatorial regions, though other factors also come into play. In an El Niño year, western Canada and parts of the northern United States generally experience a milder-than-usual winter.

LITTLE BOY
Red denotes the unusual warmth associated with El Niño. The name, given by South American fishermen, means "Little Boy," referring to the infant Jesus; El Niño's appearance coincided with Jesus' birthday at Christmas.

DRY TURNS TO WET
Northern Chile, normally arid, has heavy rains in an El Niño year. Experiments are under way to see whether, with careful livestock management, the phenomenon can be exploited to renew overgrazed lands.

WIND AND WATER

El Niño events illustrate the powerful influence of winds and ocean currents on global climate and human activities. Many of the voyages of exploration described in this book involved the exploitation of phenomena that are now understood to be parts of a worldwide system.

At the beginning of his first epoch-making voyage, Christopher Columbus sailed from Pales in Spain to the Canary Islands; only then did he set the westward course that he believed would take him straight to Japan. This proved to be an excellent decision, since Columbus had unwittingly placed his ships in the path of the northeasterly trade winds, which carried them right across the Atlantic to the Bahamas. On the return journey, Columbus experienced some difficulties before he located the westerly winds needed to bring the *Niña* and the *Pinta* home by a more northerly route.

For many other explorers during the great age of sail, from Bartholomeu Dias to James Cook, success or failure hinged on what could be known or guessed about the winds and ocean currents. Experience of the way they operated accumulated over time, but their working as a global system began to be understood only in the twentieth century, and some features remain mysterious even today.

Many factors influence local wind conditions, but the major system of prevailing winds results from the interaction of two forces: the heat of the Sun and the spinning of the Earth. The Sun warms the atmosphere unevenly, so that hot air at the equator rises and cold air from other regions flows in to replace it. The movements of hot and cold air between the equator and the poles would form a simple north–south wind pattern but for the intervention of other influences, of which the most important is the Earth's rotation. This skews the winds around so that they blow toward the east where the wind is moving away from the equator, and toward the west where it blows toward the equator.

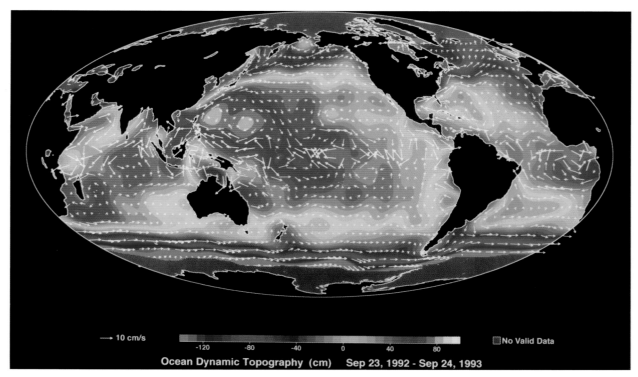

Above: This false-color map combines information on the height of sea level with indications of the direction of surface currents.

Right: Waves batter the wrecked pier at Santa Monica, California, during an El Niño storm in 1983. In El Niño years the West Coast receives more storms and rainfall than normal.

The result can be represented (in highly simplified terms) as a sphere divided into horizontal bands or zones, with the doldrums, a calm belt with only light, variable winds, at the center along the equator. The zones above and below the equator are roughly symmetrical in layout, with the trade winds, westerlies, and polar easterlies succeeding one another. Between the trade winds and the easterlies (at roughly 30 degrees north and south) lie two other calm areas, the horse latitudes, formed by descending equatorial air.

The constant motion in the atmosphere is paralleled by that of the seas and oceans. The link between them is close, since surface currents are wind driven. Currents flow like great rivers within the oceans, owing to the fact that warm and cold waters tend to remain separate (warm water rises, while cold and/or saline water is denser and sinks). The wind-driven currents are on or close to the surface, usually reaching a depth of between 330 and 660 ft. (100 and 200 m). They move west or east according to the winds, but may be deflected by other currents and forced north or south by the presence of landmasses, so that in the open ocean the currents move in huge circles, or gyres, clockwise in the northern hemisphere and counterclockwise in the south.

The Gulf Stream

The best-known gyres are found in the North Atlantic and North Pacific. The Gulf Stream is actually the northern phase of the North Atlantic gyre, driven north from the Caribbean by the trade winds and then deflected northeastward by the Labrador Current. The Gulf Stream's waters are generally held to be responsible for the climate of northwest Europe, which is much warmer than other regions at similar latitudes. In the North Pacific the Japan, or Kuroshio, Current has a similar effect in its movement from the South China Sea to Japan.

El Niños are only vivid and visible examples of the global effects of winds and ocean currents. And apart from those driven by winds, there are deep-sea currents that move much more slowly but shift vast quantities of water as they sink to the bottom in the polar regions and flow slowly toward the equator. The process is still not fully understood, but its importance is beyond doubt.

Above: On the other side of the world from Santa Monica, animal remains in New South Wales reveal the effects of a drought in Australia, another El Niño-linked phenomenon.

SAN FRANCISCO BAY
MAPPING LAND USE

Since its inception in 1972, the Landsat satellite program has produced images of the Earth, created by remote sensing, that record features such as land use with a new precision. This image of the San Francisco Bay Area derives from data accumulated in September 1997 by the Landsat 5 Thematic Mapper.

The sensors aboard the satellite record electromagnetic energy reflected off Earth, including invisible infrared waves. This allows the creation of a combined infrared and visible image, or color composite, as here. Dark gray shows the concrete and asphalt that signify dense habitation, pink to red hues are areas of rock and exposed land, green areas indicate vegetation, and water appears as dark blue or lighter shades for shallows. Similar maps can be made employing digital elevation data to achieve a three-dimensional effect, and the technology is becoming more refined. Landsat mapping of San Francisco Bay highlights the semiaridity of the region, particularly in this fall image. Its geological character is revealed by the presence of faults (notably the San Andreas Fault) and by breaks in the pattern of vegetation. Above all, such images monitor urban development, which has encircled the bay and spread south, with serious implications for agriculture, transport, and the environment.

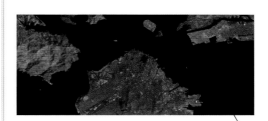

DOWNTOWN AREA
Downtown San Francisco is located at the head of a peninsula, linked by much-celebrated bridges to Marin County and the East Bay area. The tiny island is the notorious former prison of Alcatraz.

SALT AND SURF
Salt pans show up around the southern end of the bay. Shallow water is indicated by light blue, which also appears along the coastline, where it represents surf caught at a moment in time.

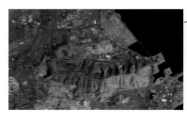

BARE MOUNTAINS
The San Bruno mountains, rising up south of downtown San Francisco. The isolated patch of hot red indicates their bareness and their independence of the mass of human habitation surrounding them.

FAULT LINE
These reservoirs are visibly part of a single geological fault line that passes just west of San Francisco. Showing up at other places on and off the map, this is the notorious and seismically threatening San Andreas Fault.

MAPPING THE PLANET

In modern times cartography has been revolutionized by technological progress. Orbiting satellites capture images of the Earth's surface and transmit data in quantities that would once have been unmanageable, but which can now be stored, synthesized, and manipulated by powerful computers. The information and understanding acquired through such tools may even help to save the planet.

As in many fields, research during World War II laid the groundwork for breakthroughs made in the 1950s. The first assembly-line computers were built, and in 1957 the Soviet Union put an artificial satellite, Sputnik 1, into orbit round the Earth. From that time onward, computers became ever more complex, capacious, and fast, while communications satellites became the nodal points of global television and telephone networks.

Computers began to make an impact on cartography in the 1960s, when the first geographical information systems (GIS) were introduced. The earliest large-scale example of such a system was designed in 1966 for the Canadian government and was employed to manage forestry and other types of land use. Essentially, GIS integrated computer hardware and geographical data with software systems whose programs could store, analyze, and manipulate the data to create automated maps. As a result, computer maps have been drawn on a scale, and of a detail and complexity, that could never have been accomplished by traditional methods. Fewer and fewer maps are now drawn by hand: even where the task is not insuperable, speed and economics favor the computer.

The Landsat Program
The creation of software capable of dealing with vast quantities of information was timely, for images and data encompassing the globe poured in as soon as satellites were equipped with remote-sensing scanners. The longest-running and most important series of Earth-monitoring satellites developed from ideas formulated in the 1960s by the U.S. Department of

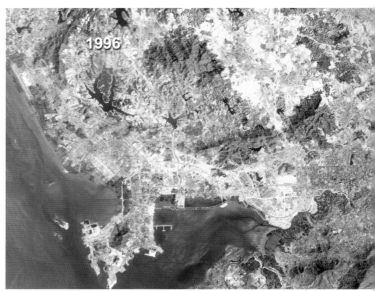

Above: Two Landsat images, taken eight years apart, show the growth of the Chinese city of Shenzhen. Settlement has spread across the reds and browns that mark barren or rocky land.

the Interior. In the 1970s these took shape as the Landsat Program (originally the Earth Resources Satellite Program), in which satellites were launched by the National Aeronautics and Space Administration (NASA) while the United States Geological Survey (USGS) organized the archiving of data and the distribution of products derived from it. Fundamental to the concept of the program was that the satellites would not only record the state of the Earth, but would pass repeatedly over each area, registering the changes occurring everywhere on the planet.

Above: Two images taken from a series in which NASA scientists have combined images from various satellites, including Landsat, to zoom in on the city of Chicago, Illinois, focusing on the downtown area around the Sears Tower.

ERTS-1, later renamed "Landsat 1," was launched in July 1972. It orbited the Earth fourteen times a day at a height of 570 miles (920 km), its multispectral scanner making thirty million observations in the course of recording each 115-square-mile (298 sq. km) scene. Landsat 1's performance was far better than expected: Though its predicted life expectancy was a year, it continued to function until early 1978.

Landsats 2, 3, 4, and 5 followed in 1975–1984. Landsat 6, launched in October 1993, was the program's first major setback: It was destroyed after the rocket's upper stage failed to fire. The satellite, commercially built and managed, was part of a generally unhappy private-sector involvement begun in 1985 as a response to spiraling costs. Eventually, this led to the abandonment of continuous operation, and for a time Landsat imagery was collected only when there was a direct commercial demand for it. The disadvantages of losing continuity of data were recognized by Congress in 1992, when government supervision began to be restored. NASA again took the leading role, and a policy was adopted of making data as widely and cheaply available as possible.

Though the equipment on Landsat 1 was impressive, later satellites benefited from further technological advances. Landsat 7, launched in 1999, carries the Enhanced Thematic Mapper Plus, which provides unprecedentedly high-resolution observations, making possible even more refined analysis of land-surface processes.

Monitoring the planet
Landsat data have already had an extraordinary range of applications, from mapping difficult terrain such as the Amazon Basin to calculating the percentage of roof insulation in an urban district. The advantages for geography, geology, oceanography, forestry, and many other disciplines have already been felt. Moreover, in an era of ecological degradation and climate change, continuous monitoring of the planet is likely to prove vital in giving early warning of imminent catastrophes and prompting political leaders to make difficult but necessary decisions.

However, satellites are not the only space tools used in modern cartography. In February 2000 the space shuttle *Endeavor* completed an eleven-day mission, scanning the Earth's surface with radar signals. The data collected are being used to create a topographic map of the planet that will be the most detailed ever produced.

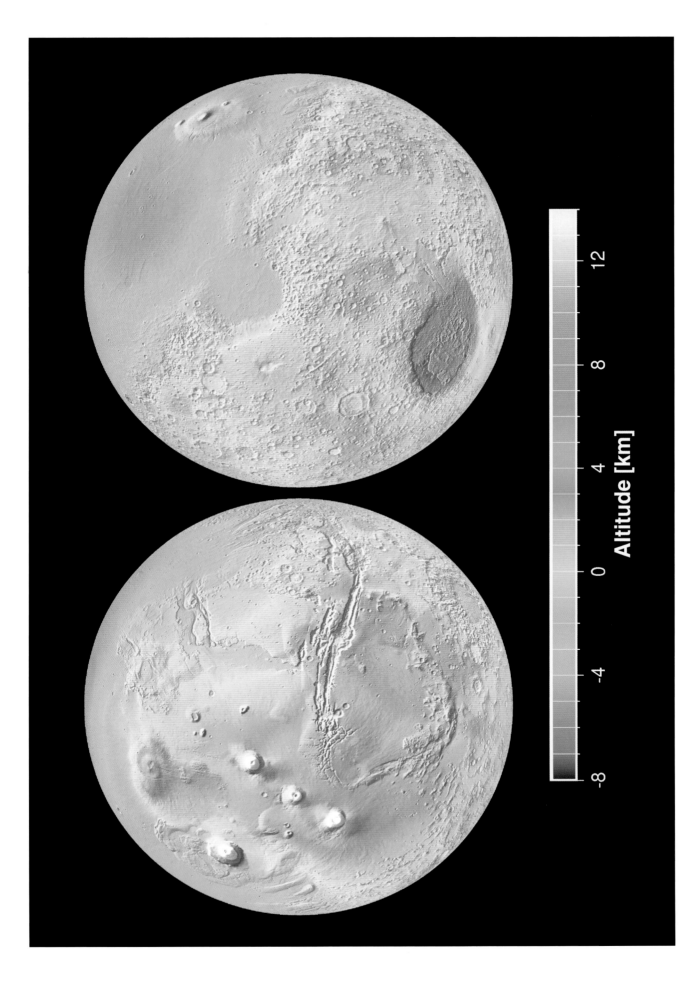

Altitude [km]

TOPOGRAPHY OF MARS
MARS GLOBAL SURVEYOR

The Red Planet, named after the Roman god of war, Mars, has been mapped and studied in many ways. In this image, false color and information from the Mars Orbiter Laser Altimeter on board the Mars Global Surveyor have been employed to create a striking topographical map, which highlights the enormous scale of some of Mars's physical features.

Maps of Mars once showed little more than contrasts of light and dark ("vegetation") on its surface. This topographical map serves to emphasize the gigantic scale of certain physical features, which are all the more dramatic in view of Mars's relatively small size (its diameter is roughly half the length of Earth's). Bombardment by meteorites has left innumerable craters on the surface, hundreds of them large and distinct enough to have been named. There is a 2,500-mile (4,000-km) long fracture in Mars' crust, the Valles Marineris, which is in places 75 miles (120 km) wide. Most impressive of all are the chains of extinct volcanoes, all in the northern hemisphere. Three are aligned northeast–southwest on a high plateau (Tharsis), at the edge of which stands the mighty Olympus Mons, three times the height of Mount Everest. A smaller range is on a plain known as Elysium. Mars is divided, roughly north–south, between low, relatively smooth areas and much more extensive, higher, and densely cratered terrain.

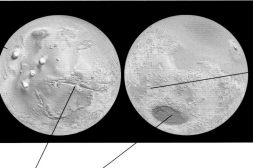

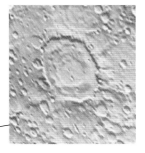

MIGHTY VOLCANOES
Olympus Mons is the largest volcano in the solar system, rising to a height of 16.4 miles (26.4 km) above the surface. To the southeast stands a row of three more gigantic volcanoes, Ascraeus, Pavonis, and Arsia Mons.

HUYGENS CRATER
The Huygens crater, the second largest on Mars, is 300 miles (495 km) across. It was named after the Dutch scientist Christiaan Huygens, who designed much improved telescope lenses and in 1659 mapped Syrtis Major.

LOW POINT
Hellas, a basin formed by an immense meteorite impact, is the lowest surface region on Mars. Its often exceptional brightness meant that for a long time astronomers mistook it for a snow-covered plateau.

GIGANTIC VALLEY
A gigantic valley, the Valles Marineris ("Mariner's Valley"), was named after the orbiter Mariner 9, which first identified it. It is actually a huge crack in the surface of the planet.

EXPLORING MARS

Perceived as the most Earthlike of the planets, Mars has fascinated astronomers for centuries, and the possibility that it might harbor some form of life has caused periodic excitement. From the 1960s information provided by probes and orbiters discredited many theories, but important questions remain unanswered.

Mars was known in ancient times, and observations of it were recorded by Greek thinkers, including Aristotle and Ptolemy. In 1610 Galileo became the first astronomer to study the planet through a telescope and detect its phases. Christiaan Huygens identified the first feature on the surface, the Syrtis Major, in 1659.

Further observations recorded, among other things, Mars's rotation and the existence of polar caps, but the fuzziness of telescopic images made it impossible to distinguish between physical and purely optical phenomena and gave rise to widely different theories. In 1877 the Italian astronomer Giovanni Virginio Schiaparelli believed that he had observed a network of canals (and, implicitly, intelligent life). Though never commanding universal consent, the theory had a long life. More widely accepted were the identification of large, dark patches as seas, and later as vegetation, and the conviction that the planet's atmosphere was mostly nitrogen.

The era of space flight put such beliefs to the test. From 1962 missions were launched by the United States and Soviet and post-Soviet Russia, but the failure rate was high. On the first successful mission in 1965, the United States' Mariner 4 sent back twenty-two pictures as it passed Mars, which revealed an arid, rocky landscape and confirmed that the planet's surface was much cratered.

In 1971–1972 Mariner 9 became the first Mars orbiter, sending thousands of pictures back to Earth. It demonstrated that Mars had seasons, recorded the

Above: An orthographic view of Mars made up from images from the Viking probes, with the north polar cap at the top and the Valles Marineris just below the equator.

Right: A false-color map of Martian topography—from blue lowlands to red and brown highlands—shows the impact craters that pock the southern hemisphere.

Left: The view from the Pathfinder lander. The Sojourner rover is visible beside the large rock (center) dubbed Barnacle Bill.

advance and retreat of the polar caps, and provided evidence to suggest that there had been liquid water on the planet in the distant past. But Schiaparelli's canals were shown to be an illusion, and there was no trace of vegetation.

Probing the theories

In 1976 two U.S. probes, Viking 1 and Viking 2, made the first landings on the surface, sending back closeup pictures and investigating the geology and atmosphere of the planet. Once more the results contradicted expectations: The atmosphere proved to be 95 percent carbon dioxide. The atmospheric pressure was too low for liquid water to exist on the surface, and the soil revealed no traces of organic compounds. So it appeared that Mars was unable to sustain life, though it might have done so in the past.

Further investigation was hindered by a series of failed missions culminating in the explosions that destroyed the United States' Mars Observer (1993) and Russia's Mars '96. The combination of huge expense and high risk led to a new "faster, cheaper" NASA strategy. In July 1997 the Pathfinder probe parachuted to the surface, where its lander and a wheeled "rover," Sojourner, began work. Two months later the even more scientifically sophisticated Mars Global Surveyor began orbiting the planet. Equipped with remote sensing devices, Surveyor compiled a year-round (that is, 688-day) record of Mars' weather, topography, gravity, and magnetism, while its camera sent back tens of thousands of images, sharper and more detailed than anything taken before.

Life and water: unanswered questions

However, interpreting the results was not easy, especially where the key issues of the presence of water and the possibility of life were concerned. In 1996 a meteorite from Mars, found much earlier in Antarctica, was claimed to contain fossils. Although this was much disputed, it raised the old issues again, reinforced by discoveries made on Earth showing that life could thrive under much more extreme conditions than had previously been believed possible.

Surveyor's contributions were ambiguous. On the one hand, its photographs showed channels in which material appeared to have been deposited in relatively recent times, a layered planetary upper crust that suggested a watery past, and, in 2001, colossal—oceanic?—channels and possible evidence of ice at the equator just a few million years ago. On the other hand, Surveyor failed to find the minerals usually formed in the presence of water, and vestigial magnetic features suggested that after the demise of Mars' magnetic field, a solar wind had blown away the nitrogen, carbon, and water in its atmosphere, inhibiting the development of life. Further mission failures delayed attempts to resolve such problems, but hopes were raised in October 2001, when Odyssey reached and began to orbit the planet.

Above: The Sojourner rover on the Martian surface. The vehicle was able to analyze the properties of the soil by applying varying amounts of pressure to reveal soil beneath the surface.

Elev 797	Var 3°W	TA 18000	TRL ATC	I-FVJ 111·15 Ch48Y	31 MAY 01	M5M

INDIANAPOLIS APPROACH 119·3	INDY TOWER 120·9	GROUND 121·9	ATIS 124·4

KIND/IND USA

Changes: New Aerodrome

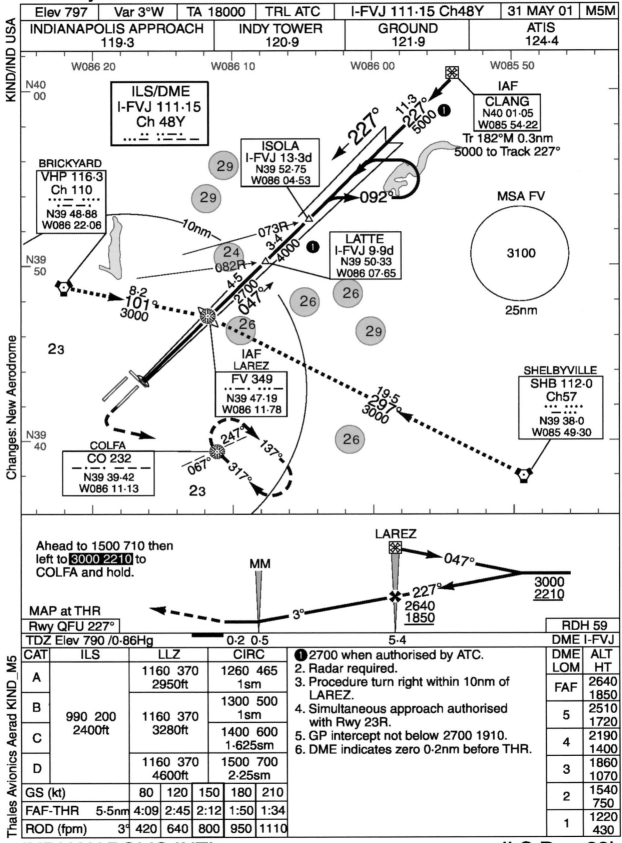

ILS/DME
I-FVJ 111·15
Ch 48Y

IAF
CLANG
N40 01·05
W085 54·22
Tr 182°M 0.3nm
5000 to Track 227°

BRICKYARD
VHP 116·3
Ch 110
N39 48·88
W086 22·06

ISOLA
I-FVJ 13·3d
N39 52·75
W086 04·53

LATTE
I-FVJ 9·9d
N39 50·33
W086 07·65

MSA FV
3100
25nm

IAF
LAREZ
FV 349
N39 47·19
W086 11·78

SHELBYVILLE
SHB 112·0
Ch57
N39 38·0
W085 49·30

COLFA
CO 232
N39 39·42
W086 11·13

Ahead to 1500 710 then left to 3000 2210 to COLFA and hold.

LAREZ

MM

047°

3000 2210

227°
2640
1850

3°

MAP at THR

| Rwy QFU 227° | | | | | | RDH 59 |
| TDZ Elev 790 /0·86Hg | | 0.2 0.5 | | 5·4 | | DME I-FVJ |

CAT	ILS	LLZ		CIRC			❶ 2700 when authorised by ATC.			DME LOM	ALT HT
A		1160 370 2950ft		1260 465 1sm			2. Radar required.			FAF	2640 1850
B	990 200 2400ft	1160 370 3280ft		1300 500 1sm			3. Procedure turn right within 10nm of LAREZ.			5	2510 1720
C				1400 600 1·625sm			4. Simultaneous approach authorised with Rwy 23R.			4	2190 1400
D		1160 370 4600ft		1500 700 2·25sm			5. GP intercept not below 2700 1910.			3	1860 1070
GS (kt)		80	120	150	180	210	6. DME indicates zero 0·2nm before THR.			2	1540 750
FAF-THR 5·5nm		4:09	2:45	2:12	1:50	1:34				1	1220 430
ROD (fpm) 3°		420	640	800	950	1110					

Thales Avionics Aerad KIND_M5

INDIANAPOLIS AIRPORT
INSTRUMENT APPROACH CHART

Air travel requires maps that seem abstract and sparse compared with conventional maps of the Earth's surface. Pared of unnecessary distractions, the instrument approach chart guides pilots arriving at, or departing from, major airports, in this case Indianapolis International.

The instrument approach chart supplies a pilot with the information needed to find and land on a particular runway, using only onboard instruments and electronic airport guidance systems. The chart is used once air traffic control has told the pilot which runway to head for and has guided the plane to the initial approach fix (IAF). The large diagram is a general plan of the approach, and its most prominent feature is a bold diagonal "arrow," known as the localizer. It represents the electronic beam, running along the line of approach, that will guide the plane to the runway. At marked points along the beam, the pilot can crosscheck the plane's position and altitude. The studded lines and directional arrows leading to LAREZ indicate the course to be taken by an aircraft from the direction of Brickyard or Shelbyville. Having joined the localizer, the plane will fly away from the airport, beginning a turn on to 092 degrees after passing Isola that will align it for the approach. The smaller diagram is the vertical profile of the descent. Another electronic beam, the glideslope, ensures that it takes place at the correct angle.

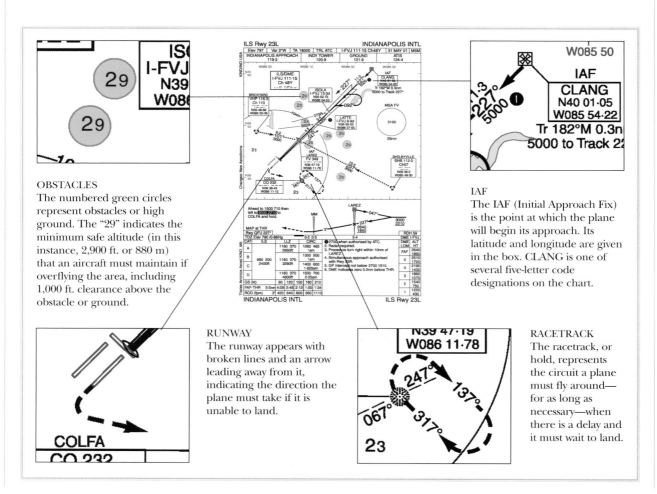

OBSTACLES
The numbered green circles represent obstacles or high ground. The "29" indicates the minimum safe altitude (in this instance, 2,900 ft. or 880 m) that an aircraft must maintain if overflying the area, including 1,000 ft. clearance above the obstacle or ground.

IAF
The IAF (Initial Approach Fix) is the point at which the plane will begin its approach. Its latitude and longitude are given in the box. CLANG is one of several five-letter code designations on the chart.

RUNWAY
The runway appears with broken lines and an arrow leading away from it, indicating the direction the plane must take if it is unable to land.

RACETRACK
The racetrack, or hold, represents the circuit a plane must fly around—for as long as necessary—when there is a delay and it must wait to land.

AIR TRANSPORT

Sophisticated navigational aids and safety
procedures are only one aspect of the
advanced technology that has become
commonplace in civil and military
aviation. Though relatively recent, the
phenomenal growth of commercial air
transport has been an important force
for change.

Mass air travel has become so familiar that it is
easy to forget that, until December 1903, the
only known forms of manned flight were by
lighter-than-air craft—balloons and airships. Then
Orville Wright spent twelve world-changing seconds
in the air at Kittyhawk, North Carolina, and aviation
entered a heroic age of adventure and public
enthusiasm. Unsurprisingly, it was an age in which the
prosaic advantages of regular and safe passenger
travel did not yet exist.

During this period, stunts and record-breaking
feats hit the headlines every few years. In 1909 Louis
Blériot made the first flight across the English
Channel (four years later he executed the first loop-
the-loop). In 1919 Alcock and Brown crossed the
Atlantic from Newfoundland to Ireland. Public
acclaim for aviators reached unprecedented heights in
1927, when Charles A. Lindbergh flew his monoplane,
The Spirit of St. Louis, from New York to Paris in $33^1/_2$
hours. And in the 1930s the exploits of women
aviators such as Amy Johnson, Amelia Earhart, and
Beryl Markham maintained the high-profile, human-
interest appeal of flying.

Meanwhile, aircraft had demonstrated their
military potential in World War I, and technological
improvements meant that they flew ever faster, higher,
and farther. Commercial air services were relatively
slow to develop, however. In 1919 regular international
flights began between London and Le Bourget
(Paris), and other routes were gradually established.
But air transport was not commercially viable, and it
depended on government assistance. Airlines were
either nationalized or given subsidies, often in the
form of contracts to deliver mail. Airmail carriers
pioneered many long-distance routes with a heroic

tenacity comparable to, though less celebrated than,
those of individual record-breaking aviators. This was
true of transcontinental airmail flights in the U.S. and
also of the services developed to link centers within
colonial empires: Paris–Dakar, Paris–Saigon, London–
Singapore–Brisbane, Cairo–Cape Town, and so on.

Early passenger services

During the 1920s and 1930s many famous airlines
were established, often as a result of amalgamations:
the Dutch KLM line, Sabena in Belgium, Britain's
Imperial Airways, Lufthansa in Germany, Pan
American, and Air France. Scheduled passenger
services were gradually set up, mostly along already
existing mail routes (whose patterns, even today,
influence some air links). Improvements in
navigational instruments and general performance
continued, and developments such as the flying boat
sustained the romantic image of aviation. However,
the number of people who could afford to travel by
air remained small and cargo capacity was limited.

Above: A different age of air travel: twin-propellor Dakotas at an
airfield in Delhi, India, in the 1930s. They would have linked
the city with other centers of the British Empire.

Left: Standiford Field, at Louisville, Kentucky, seen from the air. The growth of domestic flights in the United States means that even relatively small airports can be very busy—and their safety procedures correspondingly complicated.

Before World War II the most commercially successful civil aviation industry was that of the United States, which benefited enormously from the scale of operations possible within its huge single market: The U.S. domestic network was comparable in size to the international routes of divided Europe. Notable American achievements were the highly advanced Douglas DC-2 and DC-3 liners and the transatlantic and transpacific routes pioneered by PanAm.

World War II brought a host of technological advances, including the first jet aircraft. But in the 1950s, British attempts to begin a jet passenger aircraft service were beset by technical difficulties, and the jet age in civil aviation began only in 1958, when the Boeing 707 and the DC-8 entered service. At about the same time, steadily increasing prosperity in the most advanced economies fostered mass tourism. A large new market was quickly catered for by cheap flights and "package" deals in which fares and hotel accommodations were offered for an all-in fee.

As the aviation industry was deregulated, new companies sprang up, and from 1970 an ever-increasing demand was met by the introduction of wide-bodied aircraft such as the Boeing 747, capable of carrying hundreds of passengers at a time. Supersonic passenger travel was inaugurated in 1976, when the Anglo-French Concorde went into service.

Though global in reach, mass air travel is still confined to North America, Europe, the Pacific Rim, and Australasia. In these areas, with traffic increasing every year, there is no foreseeable end to the pleasures and problems brought by mass air travel.

Above: The introduction in the 1970s of wide-bodied jets, like these Boeing 747s at London's Heathrow Airport, dramatically increased passenger-carrying capacity. Yet parts of Africa, Asia, and South America still lack mass passenger flight networks.

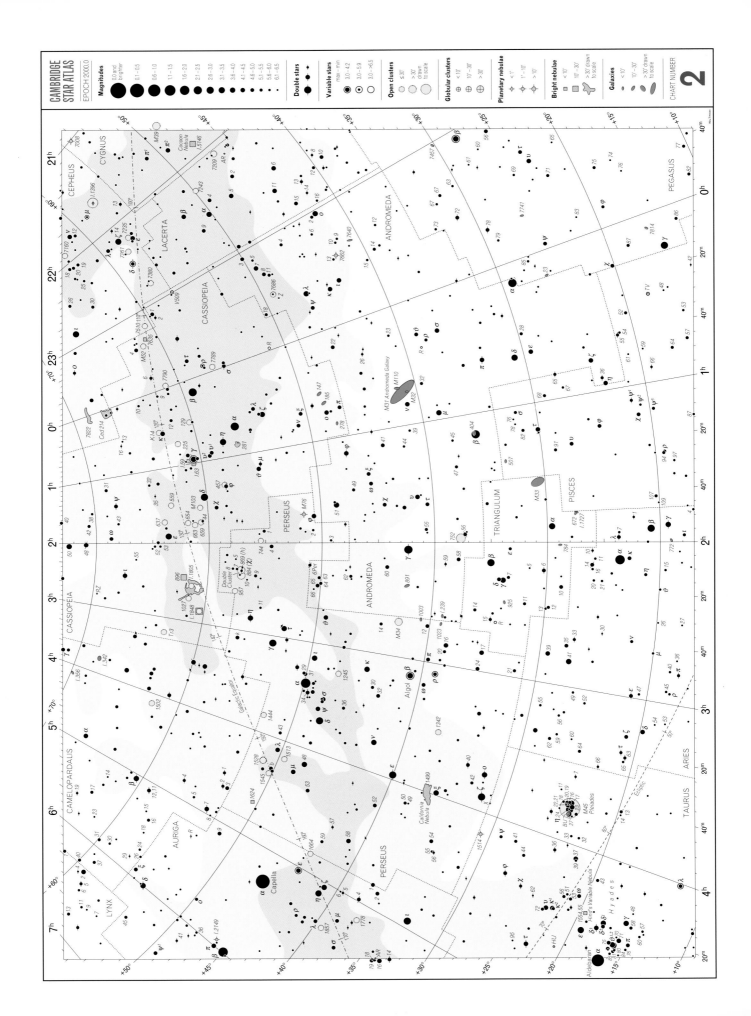

STAR CHART, 2001
WIL TIRION

This chart from the *Cambridge Star Atlas* represents just one sector of the sky, crowded with stars, clusters, nebulae, and galaxies. Nevertheless, the display only hints at the wonders of the heavens, for the stars shown are restricted to those that can be seen with the naked eye or with the help of a moderately powerful telescope.

Star charts are designed as though the heavenly bodies lie on the inside surface of an enormous, transparent sphere. The sky appears as it would look to an observer who was unaware that the objects lay at different distances from the Earth. Like the Earth, the celestial sphere has poles, an equator, and an imaginary grid from which readings can determine the position of any object. Instead of longitude and latitude, the star chart is crossed by celestial coordinates known as right ascension, reckoned in hours, minutes, and seconds, and declination, reckoned in degrees, minutes, and seconds. Various projections can be employed to represent the celestial sphere on a flat surface. Here the secant conical projection is used, in which the right ascension appears as north–south straight lines and the lines of equal declination are evenly spaced. The boundaries of constellations are marked by dotted lines, as are the Ecliptic (bottom left, the apparent path of the Sun among the stars) and the Galactic Equator running across the top.

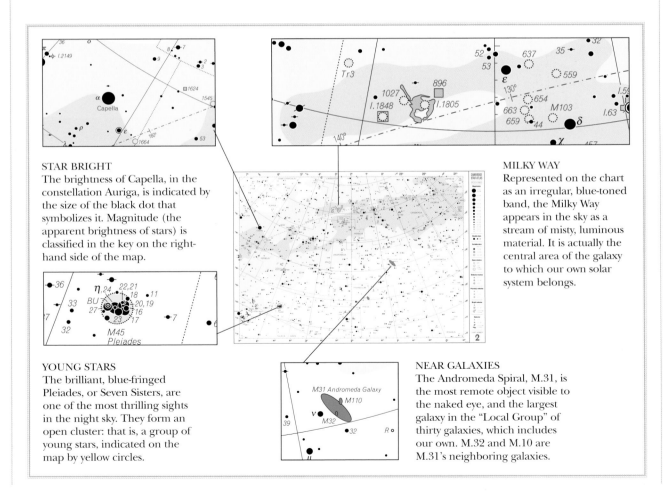

STAR BRIGHT
The brightness of Capella, in the constellation Auriga, is indicated by the size of the black dot that symbolizes it. Magnitude (the apparent brightness of stars) is classified in the key on the right-hand side of the map.

YOUNG STARS
The brilliant, blue-fringed Pleiades, or Seven Sisters, are one of the most thrilling sights in the night sky. They form an open cluster: that is, a group of young stars, indicated on the map by yellow circles.

MILKY WAY
Represented on the chart as an irregular, blue-toned band, the Milky Way appears in the sky as a stream of misty, luminous material. It is actually the central area of the galaxy to which our own solar system belongs.

NEAR GALAXIES
The Andromeda Spiral, M.31, is the most remote object visible to the naked eye, and the largest galaxy in the "Local Group" of thirty galaxies, which includes our own. M.32 and M.10 are M.31's neighboring galaxies.

CAMERA ON THE UNIVERSE

Large Earth-based telescopes have scanned the heavens, making major discoveries about the nature of the universe. Men have stood on the Moon, and probes have ranged through the solar system. The most breathtaking celestial images have been produced through a combination of scientific imagination and technological expertise by the Earth-orbiting Hubble Space Telescope.

The telescope is named in honor of Edwin P. Hubble (1889–1953), probably the greatest astronomer of the twentieth century. The naming was entirely appropriate in view of Hubble's achievements. Working at the Mount Wilson observatory in California, he made two discoveries that have shaped our conception of the universe: that the Milky Way is only one of many galaxies; and that, since the galaxies are moving away from us at increasing speeds, the universe must be expanding.

Images of the universe obtained from terrestrial telescopes, however informative or beautiful, suffer some distortion because they must pass through Earth's atmosphere. The concept of a large telescope operating from an orbiting space station was put forward during Hubble's lifetime, but serious work on it was begun by NASA only in the 1970s. The telescope was built from 1979 and scheduled to be launched in 1986, but all space shuttles were grounded after the *Challenger* disaster in January 1986. Finally, in April 1990, the space shuttle *Discovery* delivered the Hubble Space Telescope to its orbital position, at an altitude of 370 miles (600 km) above the Earth.

An embarrassing error

Shortly after "Hubble" became operational in the following May, it became apparent that the images it was recording were not correctly focused. One of the mirrors had been ground to the wrong shape—an embarrassing error to have occurred in a project costing an estimated $1.5 billion, and one that received worldwide publicity. Less remarked upon

Above: A Hubble image of NGC 4414, a spiral galaxy about 60 million light years away. Such a galaxy contains hundreds of millions of stars; older stars are clustered at the center, younger, bluer stars in the arms, where dark clouds of interstellar dust are also visible.

Above: A Hubble image based on the heat of various gases shows shock waves, formed by the explosion of a star around 15,000 years ago, as they pass through clouds of interstellar gas.

was the efficiency and determination with which preparations were made to put matters right, involving a year-long training of the seven-astronaut team designated to effect the repairs while Hubble remained in orbit. Undertaken in 1993, the mission was a complete success, and no significant problems have since arisen.

Hubble functioned in essentially the same way as large telescopes on Earth, capturing light on a large primary mirror that reflected it on to a smaller secondary mirror at the top of the tube. This in turn reflected the light back down the length of the telescope and through a hole in the center of the primary mirror to the instrument cluster at the base of the telescope. There, the appropriate camera and/or other device recorded it.

Since it was impractical to regrind the distorted primary lens, the astronauts repaired Hubble by installing a corrective unit that would refocus the light. To make room for it, one item in the cluster, the photometer (which measures the intensity of light), had to be removed, leaving two cameras and two spectrographs (which turn electromagnetic radiation or sound waves into a spectrum that can be photographed). One camera and spectrograph are

used for detailed recording of small areas; the other two record wider fields. Astronauts service Hubble every few years, and improved instruments have been installed that carry out the same tasks more efficiently across a wider range of wavelengths.

The data and images sent back by Hubble take the form of radio signals, transmitted via satellite relays to the Goddard Space Flight Center and the Space Telescope Science Institute, both in Maryland. Stored as computer data and enhanced or false-colored to increase contrast and clarity, the images have been of stunning quality. Hubble has recorded episodes in the birth and death of stars and galaxies, the spectacular explosions of the huge dying stars known as supernovae, and, in 1994, the impact on Jupiter of the fragmenting Shoemaker–Levy 9 comet. It has captured the first images of the surface of Pluto, the outermost planet in the solar system. Evidence provided by Hubble has confirmed the existence of large superplanetlike brown dwarfs; and it has also supported the hypotheses of vast, matter-consuming black holes and of the "Big Bang," an explosion of matter that is believed to have caused the ongoing expansion of the universe discerned by Hubble.

Hubble is expected to continue functioning until 2010. By 2008 its replacement, Next Generation Space Telescope, should be in place, ready to probe the heavens even more deeply and to continue the task of mapping the universe.

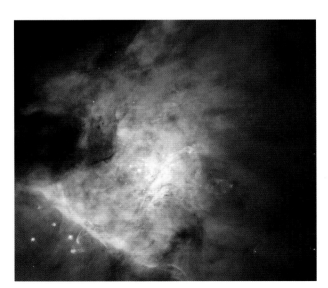

Above: A map of our own creation? Astronomers believe that disks visible in this Hubble image of the Orion nebula are embryonic solar systems, forming in the same way ours did.

MANHATTAN SNAPSHOT
THE NYCMAP

This aerial photograph shows the southeastern end of Central Park and the upper part of Midtown Manhattan, the location of the city's most fashionable and luxurious skyscrapers, stores, and offices. Remarkably distinct in itself, the picture is only one small segment of an urban mapping project unprecedented in its scope and complexity.

During 1996 the whole of New York City was photographed from the air. The digitized results—1,672 overhead views—gave images of such high resolution that it was possible to see any ground-level object a foot (30.5 cm) or more in length. The images were the base for the most comprehensive and detailed city map ever made, the "NYCMap." The mapmakers first had to convert the photographs into a planometric map. The basic version of this familiar type of city map appears overleaf, showing the same area as here; but the full NYCMap was created by identifying and naming millions of features on the photographs, from streets and buildings to individual sewers and phone booths. Consequently the NYCMap really comprises a gigantic single resource from which an almost infinite number of maps can be made according to the requirements specified by a particular user. The map was finished in 2000–2001—in so far as it can ever be finished, since it will be updated for the foreseeable future.

TWIN TOWERS
The twin towers of Century Apartments are a well-known Manhattan landmark, still eye-catching when viewed from the air. Like much of the West Central Park district, they are the haunts of celebrities.

RINK
Amphitheater-like from the air, Wollman Memorial Rink offers the harmless pleasures of ice skating in winter and roller skating and minigolf in summer. To its right, spacious Central Park Zoo; below, the modestly named Pond.

TEMPLE OF ART
New York's great Museum of Modern Art (MoMA) takes up much of the block between 5th Avenue and the Avenue of the Americas. The large, leafy sculpture garden on West 54th Street is clearly visible.

THROUGH TRAFFIC
Beads of colour dot the grid representing the streets and avenues of Manhattan as automobiles pass—or crawl—along them. In this detail they are in front of the well-known Trump Tower on East 57th Street.

NEW YORK, NEW YORK

The vast metropolis recorded by the NYCMap is an extraordinary historical phenomenon—from little Dutch settlement to cosmopolitan colossus in under four hundred years. Now effectively the financial and commercial capital of the world, New York has a restless, dynamic character unlike that of any other city.

At the time of the American Revolution, New York had a population of 25,000 and was still confined to the southern tip of Manhattan Island. Divided in its allegiance, the city was occupied by the British from 1776 to the end of the war, during which it served as their military headquarters and provided a refuge for loyalists.

Despite this inauspicious episode, New York became the capital of the USA in 1785, and George Washington was sworn in as first president in the city's Federal Hall. But in 1790 the capital was transferred to Philadelphia and six years later New York was replaced as the state capital by Albany.

Ironically, when these demotions occurred, New York's rise to national preeminence had already begun. The city's seaboard location and superb harbor made it a magnet for domestic and foreign trade, which in turn promoted the growth of financial and insurance services. The completion of the Erie Canal in 1825 linked New York with the West, and at about that time the city overtook Philadelphia and Boston to become the largest in the United States.

In the 1840s and 1850s its population rapidly increased as Irish, German, and other immigrants arrived. Tensions between native- and foreign-born New Yorkers led to a number of riots, but the immigrant vote became a significant political force when it was mobilized by Tammany Hall. Originally a charitable institution, Tammany became the Democratic Party machine, controlling New York for the best part of a century; despite widespread corruption and sensational trials such as that of "Boss" Tweed in the early 1870s, reform candidates enjoyed only limited success.

After the Civil War, New York entered the "Gilded Age" of conspicuous consumption by the few, embodied in figures such as Andrew Carnegie, John D. Rockefeller, and John Jacob Astor. Dramatic scenes occurred on Wall Street as booms alternated with panics. Private opulence was matched by the building of the Waldorf Astoria Hotel, the Metropolitan Opera,

Below: The basic planometric map created from the aerial survey is just the first and most traditional level of the NYCMap. Other layers show high degrees of detail—down to sewer outlets and manholes—that can be used to produced customized maps of the city of an unparalleled accuracy.

Left: A proud symbol of New York's grandeur at the end of the nineteenth century, the Brooklyn Bridge was celebrated as an engineering marvel unmatched in any other city.

and other landmarks, while Broadway began its career as theaterland. There were also new amenities for the less affluent. Land acquired in 1856 was transformed over two decades into Central Park. Bloomingdale's and Macy's opened, the Elevated Railway was constructed, the first subway began operations in 1904, and Grand Central Station was completed in 1913. In contrast, new immigrants from Eastern Europe and Italy lived in appalling conditions, and for a time the Lower East Side is said to have been the most densely inhabited place on earth.

With the opening of the Brooklyn Bridge in 1883, the first major physical link was forged between Manhattan and the surrounding boroughs. In 1898 the five boroughs (Manhattan, the Bronx, Brooklyn, Queens, and Staten Island) were merged to form Greater New York, the second largest city in the world. Expansion continued as New York shared the nation's experience of World War I, Prohibition, and the Jazz Age, which was also the skyscraper age, culminating in the erection of the (then) world's highest, the Empire State Building.

The city's tragedy
New York emerged from World War II as the financial capital of the world and the greatest city of the most powerful nation. Physically it underwent enormous changes; long-serving City Parks Commissioner Robert Moses left a legacy of developments—housing, highways, dams, bridges—often seen as a mixed blessing. The city's late-twentieth-century history was a rollercoaster affair, with bankruptcy threatened in the 1970s, the 1980s boom collapsing with a crash in 1987, the "information age" gathering speed in the 1990s, and terrorist alarms culminating in the catastrophe of September 2001, when terrorists destroyed the World Trade Center at the cost of thousands of lives.

In addition, New York was faced with all the modern urban problems—pollution, congestion, a deprived underclass, drugs, crime—in accentuated form. Immigration made the city more diverse than ever, with a prewar surge in the black population and postwar influxes of Puerto Ricans and fellow Hispanics, Koreans, Jamaicans, and many others. Such ethnic diversity has its problems, but it also gives New York much of the energy, variety, and color that continue to characterize it.

Above: Manhattan at the start of the twenty-first century: an iconic image of towering skyscrapers and canyonlike streets.

Further Reading

Aczel, Amir D. *The Riddle of the Compass.* Harcourt Brace, 2001.

Allen, P. *The Atlas of Atlases.* Harry N. Abrams, 1992.

Bagrow, L., and R.A. Skelton. *History of Cartography.* Watts, 1964.

Barber, P. *Tales from the Map Room.* BBC Books, 1993.

Berthon, S., and A. Robinson. *The Shape of the World.* George Philip, 1991.

Black, Jeremy. *Maps and History: Constructing Images of the Past.* Yale University Press, 1997.

Booth, J. *Looking at Old Maps.* Cambridge House, 1979.

Brotton, Jerry. *Trading Territories: Mapping the Early Modern World.* Cornell University Press, 1998.

Brown, Lloyd Arnold. *The Story of Maps.* Dover Publications, 1979.

Goss, J. *Braun and Hogenberg's City Maps of Europe.* Studio Editions, 1991.

Goss, J. (ed.). *Blaeu's 'The Grand Atlas.'* Studio, 1997.

Harley, J.B., and D. Woodward (eds.). *The History of Cartography, Vol I.* University of
Chicago Press, 1987 (and subsequent volumes; still in progress).

Harvey, Miles. *The Island of Lost Maps: A True Story of Cartographic Crime.* Random House, 2000.

Harvey, P. D. A. *Maps in Tudor England.* Public Record Office, British Library, 1993.

Harvey, P. D. A. *Medieval Maps.* British Library, 1991.

Hindle, P. *Maps for Historians.* Phillimore, 1998.

Keay, John. *The Great Arc: The Dramatic Tale of How India Was Mapped and Everest
Was Named.* HarperCollins, 2000.

Lister, R. *Antique Maps and Their Cartographers.* Bell, 1970.

Manasek, Francis J. *Collecting Old Maps.* Terra Nova Press, 1998.

Moreland, Carl, and David Bannister. *Antique Maps.* Phaidon Press Inc., 1993.

Nebenzhal, K. *Maps from the Age of Discovery.* Times Books, 1990.

Owen, T., and E. Pil. *Ordnance Survey: Mapmakers to Britain since 1791.* HMSO, 1992.

Potter, Jonathan. *Collecting Antiques Maps.* Studio Editions, 1992.

Saxton, Christopher. *Christopher Saxton's 16th Century Maps: The Counties of England and Wales.* Chatsworth Library, 1992.

Seymour, W. A. (ed.). *A History of the Ordnance Survey.* Dawson, 1980.

Tooley, R. V. *Maps and Map-Makers.* Batsford, 1987.

Tooley, R. V., and C. Bricker. *Landmarks of Mapmaking.* Wordsworth, 1989.

Whitfield, Peter. *New Found Lands: Maps in the History of Exploration.* Routledge, 1998.

Whitfield, Peter. *The Charting of the Oceans: Ten Centuries of Maritime Maps.* Pomegranate, 1996.

Wilford, John Noble. *The Mapmakers.* Knopf, 2000.

Winchster, Simon. *The Map That Changed the World: William Smith and the Birth of
Modern Geology.* HarperCollins, 2001.

WEBSITES

MAP FORUM

www.mapforum.com

BRITISH LIBRARY MAP COLLECTIONS

http://www.bl.uk/collections/maps

BODLEIAN LIBRARY MAP ROOM

http://www.bodley.ox.ac.uk/guides/maps/

PERRY-CASTAÑEDA LIBRARY MAP COLLECTION, UNIVERSITY OF TEXAS ONLINE

http://www.lib.utexas.edu/maps/index/html

THE DAVID RUMSEY COLLECTION

http://davidrumsey.com

THE HISTORY OF CARTOGRAPHY PROJECT

http://feature.geography.wisc.edu/histcart/

INDEX

Picture Credits